AF352822

Fatigue and Fracture Behaviour of Additively Manufactured Mechanical Components

Fatigue and Fracture Behaviour of Additively Manufactured Mechanical Components

Editors

Roberto Citarella
Paulo M. S. T. de Castro
Angelo Maligno

MDPI • Basel • Beijing • Wuhan • Barcelona • Belgrade • Manchester • Tokyo • Cluj • Tianjin

Editors

Roberto Citarella
University of Salerno
Italy

Paulo M. S. T. de Castro
Universidade do Porto
Portugal

Angelo Maligno
University of Derby
UK

Editorial Office
MDPI
St. Alban-Anlage 66
4052 Basel, Switzerland

This is a reprint of articles from the Special Issue published online in the open access journal *Applied Sciences* (ISSN 2076-3417) (available at: https://www.mdpi.com/journal/applsci/special_issues/Fatigue_Fracture_Additive_Manufacturing).

For citation purposes, cite each article independently as indicated on the article page online and as indicated below:

LastName, A.A.; LastName, B.B.; LastName, C.C. Article Title. *Journal Name* **Year**, *Article Number*, Page Range.

ISBN 978-3-03943-665-1 (Hbk)
ISBN 978-3-03943-666-8 (PDF)

Contents

About the Editors

Roberto Citarella obtained his Doctor degree and Ph.D. in Mechanical Engineering in 1994 and in 1999, respectively, both at University of Naples, IT. In 1996, he received his diploma for a Master's in Business Administration (MBA) at the school "Stoà" in Naples. From 2005 to 2015, he was an assistant professor for teaching the discipline of Machine Design at University of Salerno. Since March 2015, he has been an associate professor of Machine Design at the Department of Industrial Engineering (DIIN) at the University of Salerno. He was visiting researcher at the Wessex Institute of Technology in 1996, Southampton (UK), and at the Queen Mary and Westfield College, London (UK), in 1998 and 2000. He was a member of: International Scientific Advisory Committee for the international conference "Fatigue Damage of Materials 2003"; the referee group for the first European call FP7 Transport–Aeronautics; and the organization committee for the international conference BeTeq 2007. He was involved as the main investigator and member in several national research activities. He collaborates also with international research centers such as the Max Planck Institute of Greiswald, Germany; and the Kazan Scientific Center Russian Academy of Sciences, Kazan, Russia, among others. He is a fellow of the Wessex Institute of Technology. His main topics of research are: boundary element method; vibrational acoustics; bioengineering; fracture mechanics; and thermomechanical fatigue. He has published nearly 110 technical papers in international peer-reviewed journals and conference proceedings. He serves as a reviewer for many international journals and is a member of the editorial board of the journals *"Advances in Engineering Software" and "The Open Mechanical Engineering Journal"* (for the latter, he has just been appointed regional editor).

Paulo M. S. T. de Castro obtained a first degree in Mechanical Engineering from the Faculty of Engineering of the University of Porto (FEUP) in 1973, a Master's degree from Imperial College London in 1977, and a Ph.D. from the Cranfield Institute of Technology in 1980. He is currently a retired full professor of the Department of Mechanical Engineering of FEUP. His research interests are mainly in the field of fatigue, fracture and structural integrity. A substantial part of his research in recent years has been related to aeronautics, formerly riveted structures and more recently, integral structures, particularly those made with FSW. He has been involved in several scientific and professional associations, such as ASME, where he was a member of the Board on Professional Practice and Ethics, EASN, TWI, IOM3, SEFI and Ordem dos Engenheiros, among others. He is a corresponding member of the Lisbon Academy of Sciences and a member of the editorial board of the journals *"Fatigue and Fracture of Engineering Materials and Structures"*, *"International Journal of Structural Integrity"*, *"Mechanika"*, and *"UPB Scientific Bulletin Series D: Mechanical Engineering"*, among others. He has taken sabbaticals as a Fulbright scholar at the University of California at Berkeley, and as a visiting scholar at Lehigh University. He was a professeur invité of the Université des Sciences et Technologies de Lille. He has a diversified experience as an evaluator of R&D for the European Union and several national organizations, and as a monitor for EU programs or projects, including the recent large aeronautics project LOCOMACHS. He is currently a member of the Scientific Committee of the EU program Clean Sky 2.

Angelo Maligno, Ph.D., is a professor of Composite Materials at the Institute for Innovation in Sustainable Engineering (IISE), University of Derby. Dr Angelo Maligno has significant experience in the R&D analysis and the design of structural components and he has been involved in several multi-disciplinary research and industrial projects aimed at investigating the response of advanced engineering materials and structures to various types of external loading and environmental conditions, using a combination of analytical, numerical and experimental techniques. He holds a Laurea Degree in Nuclear Engineering from the University of Palermo, Italy, and completed his Ph.D. in the Mechanics of Composite Materials at the University of Nottingham. Dr Maligno has worked in the industry and consultancy sectors, where he carried out R&D activities related to the aerospace, medical, defense, and nuclear engineering sectors. Dr. Maligno joined the University of Derby in 2014. In a joint Chair role at IISE, he is leading research into the design and mechanics of advanced materials at different scales using advanced computational modeling strategies. Professor Maligno leads a team of full-time researchers and he is assisted by visiting professors from the industry (AIRBUS, GKN AEROSPACE, ROLLS ROYCE Nuclear) and members of the research office at the University.

Preface to "Fatigue and Fracture Behaviour of Additively Manufactured Mechanical Components"

This Special Issue presents the latest advances in the field of fatigue and fracture performances of additively manufactured mechanical components, including components made of traditional materials (metals, sintered steels, etc.) but undergoing complex loading conditions (multiaxial fatigue and mixed mode fracture). This Special Issue is composed of seven papers covering new insights in structural and material engineering. The advent of additive manufacturing (AM) processes applied to the fabrication of structural components creates the need for design methodologies and structural optimization approaches that take into account the specific characteristics of the process. While AM processes enable unprecedented geometrical design freedom, which can result in significant reductions of component weight (e.g., through part count reduction), they have implications in the fatigue and fracture strength due to residual stresses and microstructural features. This is due to the stress concentration effects and anisotropy that still warrant further research. The papers of this Special Issue report on numerical simulation and experimental work, or a combination of both. The application of damage and fracture mechanics concepts, the appraisal of stress concentration effects, and the consideration of residual stresses and anisotropic behavior, are tackled for a range of structural applications from biomedical engineering to aerospace components.

Roberto Citarella, Paulo M. S. T. de Castro, Angelo Maligno
Editors

Editorial

Editorial on Special Issue "Fatigue and Fracture Behaviour of Additive Manufacturing Mechanical Components"

Roberto Citarella [1,*]**, Paulo M. S. T. De Castro** [2] **and Angelo Maligno** [3]

[1]   Department of Industrial Engineering, University of Salerno, 84084 Fisciano, Italy
[2]   Department of Mechanical Engineering, Universidade do Porto, Faculdade de Engenharia, 4200-465 Porto, Portugal; ptcastro@fe.up.pt
[3]   Institute for Innovation in Sustainable Engineering, University of Derby, Derby DE 01332, UK; A.Maligno@derby.ac.uk
*   Correspondence: rcitarella@unisa.it; Tel.: +39-089-96-4111

Received: 17 February 2020; Accepted: 25 February 2020; Published: 1 March 2020

Abstract: This Special Issue presents the latest advances in the field of fatigue and fracture performances of additively manufactured mechanical components, including components made of traditional materials (metals, sintered steels, etc.) but undergoing complex loading conditions (multiaxial fatigue and mixed mode fracture). This Special Issue is composed of seven papers covering new insights in structural and material engineering. The advent of additive manufacturing (AM) processes applied to the fabrication of structural components creates the need for design methodologies and structural optimization approaches that take into account the specific characteristics of the process. While AM processes give unprecedented geometrical design freedom, which can result in significant reductions of component weight (e.g., through part count reduction), they have implications in the fatigue and fracture strength due to residual stresses and microstructural features. This is due to stress concentration effects and anisotropy that still need research. The papers of this Special Issue report on numerical simulation and experimental work, or a combination of both. The application of damage and fracture mechanics concepts, the appraisal of stress concentration effects, and the consideration of residual stresses and anisotropic behaviour are tackled for a range of structural applications from biomedical engineering to aerospace components.

Keywords: fatigue; fracture; additive manufacturing; finite element method (FEM)

Transport systems face great pressure in terms of ever increasing performance and efficiency while ensuring maximum reliability and controlling costs. Material selection, structural design, and fabrication methods play a central role among many different contributions for achieving those objectives.

The mainly used metallic materials are steel alloys, where the introduction of special alloying may substantially improve performance, while sintered steel alloys are playing an increasingly important role because of their corrosion performance.

The emergence of additive manufacturing (AM) implies that components may become simpler, reducing weight and part count, a trend that is also supported by fabrication techniques such as friction stir welding or laser beam welding, leading to integral structures.

Open problems exist in all those areas, as exemplified by the assurance of integrity and mechanical performance of AM parts. Moreover, the benefits of AM are offset to a certain extent by the poor surface finish and high residual stresses resulting from the printing process, which consequently compromise the mechanical properties of the parts, particularly their fatigue performance.

The understanding of the mechanical behaviour and its incorporation into design practice is made through structural analysis, and this subject is also of interest for this Special Issue. Computational mechanics progressed from the traditional finite element method (FEM) and dual boundary element method (DBEM) approaches to combined/hybrid and multiscale analyses that may accurately model and predict crack paths and damage within controlled computational effort.

The purpose of this Special Issue is to draw the attention of the scientific community to recent advances in modelling and optimizing the structural behaviour of advanced materials and their possible applications, while also considering non-destructive testing and evaluation.

Theoretical, numerical, and experimental contributions describing original research results and innovative concepts on materials and structures were collected.

This Special Issue includes several high-quality papers written by leading and emerging specialists in the field. Among the articles collected, a number of high-quality papers existed, which led to seven published articles. A very short description of the addressed topics, in the order of themes cited below, is presented.

When studying the performance of additively manufactured components, an important issue is related to the correct design of the specimen in fatigue testing. In the property characterization of additive manufacturing materials, mini specimens are preferred due to the specimen preparation and manufacturing cost, but mini specimens demonstrate higher fatigue strength than standard specimens due to the lower probability of material defects resulting in fatigue. In Reference [1], a novel adaptive displacement-controlled test set-up was developed for fatigue testing using mini specimens. In this study, a dual gauge section Krouse type mini specimen was designed to conduct fatigue tests on additively manufactured materials. A fully reversed bending (R = −1) fatigue test was performed on simply supported specimens. The fatigue performance of the wrought 304 and additively manufactured 304L stainless steel was compared applying a control signal monitoring (CSM) method. The test results and analyses were useful to validate the design of the specimen and the effective implementation of the test bench in the fatigue testing of additively manufactured materials.

It was proven that surface integrity alteration induced by the machining process or AM has a profound influence on the performance of a component. The different manufacturing conditions directly affect the surface state of the parts (surface texture, surface morphology, surface residual stress, etc.) and affect the final performance of the workpiece.

In particular, with reference to AM, its benefits are offset to a certain extent by the poor surface finish and high residual stresses resulting from the printing process, which consequently compromise the mechanical properties of the parts, particularly their fatigue performance. Ultrasonic impact treatment (UIT) is a surface modification process which is often used to increase the fatigue life of welds in ship hulls and steel bridges. In Reference [2], the benefits of ultrasonic impact treatment (UIT) on the fatigue life of Ti-6Al-4V, manufactured by direct metal laser sintering (DMLS), were illustrated. Results showed that UIT enhanced the fatigue life of DMLS Ti-6Al-4V parts by suppressing the surface defects originating from the DMLS process and inducing compressive residual stresses at the surface. At the adopted UIT application parameters, the treatment improved the fatigue performance by 200%, significantly decreased surface porosity, reduced the surface roughness by 69%, and imposed a compressive hydrostatic stress of 1644 MPa at the surface.

On the other hand, with reference to milling technology, which can process parts of different quality grades according to the processing conditions, it is of great significance to reveal the mapping relationship between working conditions, surface integrity, and part performance for the rational selection of cutting conditions. In Reference [3], the effects of cutting parameters such as cutting speed, feed speed, cutting depth, and tool wear on the machined surface integrity during milling were thoroughly reviewed. At the same time, the relationship between the machined surface integrity and the performance of parts was also revealed. Furthermore, problems that exist in the study of surface integrity and workpiece performance in the milling process were pointed out with the final suggestion that more research should be conducted in this area in the future.

When considering dentistry applications of newly advanced materials, one issue is related to the proper adjustment of crown implant abutment during installation.

In Reference [4], the fracture resistance and stress distribution of zirconia specimens were compared considering four occlusal surface areas of implant abutment. Four implant abutment designs with 15 zirconia prostheses over the molar area per group were prepared for cyclic loading with 5 Hz, 300 N in a servo-hydraulic testing machine until fracture or automatic stoppage after 30,000 counts. Four finite element models were simulated under vertical or oblique 10-degree loading to analyse the stress distribution and peak value of zirconia specimens. Data were statistically analysed, and fracture patterns were observed under a scanning electron microscope. Cyclic loading tests revealed that specimen breakage had moderately strong correlation with the abutment occlusal area ($r = 0.475$). Specimen breakage differed significantly among the four groups ($p = 0.001$). The lowest von Mises stress value was measured for the prosthesis with a smallest abutment occlusal surface area (SA25) and the thickest zirconia crown. Thicker zirconia specimens (SA25) had higher fracture resistance and lowest stress values under 300-N loading.

The second part of this Special Issue is concerned with traditional materials but under complex fatigue conditions, like those generated in rails and wheels undergoing rolling contact fatigue with consequent crack initiations. Such cracks then develop under non-proportional mixed mode I/II/III loading, whose assessment represents a challenge for scientists involved in railway accident prevention.

In Reference [5], fatigue tests were performed to estimate the coplanar and branch crack growth rates on rail and wheel steel under non-proportional mixed mode I/II loading cycles simulating the load on rolling contact fatigue cracks; sequential and overlapping mode I and II loadings were applied to single cracks in the specimens. Long coplanar cracks were produced under certain loading conditions. The fracture surfaces observed by scanning electron microscopy and the finite element analysis results suggested that the growth was driven mainly by in-plane shear mode (i.e., mode II) loading. Crack branching likely occurred when the degree of overlap between these mode cycles increased, indicating that such a degree of enhancement led to a relative increase in the maximum tangential stress range, based on an elasto-plastic stress field along the branch direction, compared to the maximum shear stress. Moreover, the crack growth rate decreased when the material strength increased because this made the crack tip displacements smaller. The branch crack growth rates could not be represented by a single crack growth law since the plastic zone size ahead of the crack tip increased with the shear part of the loading due to the T-stress, resulting in higher growth rates.

In Reference [6], sequential and overlapping mode I and III loading cycles were applied to single cracks in round bar specimens. The fracture surface observations and the finite element analysis results suggested that the growth of long coplanar cracks was driven mainly by mode III loading. The cracks tended to branch when increasing the material strength and/or the degree of overlap between the mode I and III loading cycles. The equivalent stress intensity factor range that could consider the crack face contact and successfully regress the crack growth rate data was proposed for the branch crack. Based on the results obtained in this study, the mechanism of long coplanar shear-mode crack growth turned out to be the same regardless of whether the main driving force was in-plane shear or out-of-plane shear.

The last paper in this Special Issue concerns real components and, in particular, aerospace structures, whose residual life in the presence of a service crack is evaluated.

In Reference [7], the authors presented the results of a systematic crack propagation analysis campaign performed on a compressor-blade-like structure. The point of novelty was that different blade design parameters were varied and explored in order to investigate how the crack propagation rate in low cycle fatigue (LCF, at R ratio $R = 0$) could be reduced. The design parameters/variables studied in this work were as follows: (1) the length of the contact surfaces between the dovetail root and the disc, and (2) their inclination angle. Effects of the friction coefficient between the disc and the blade root were also investigated. The LCF crack propagation analyses were performed by recalculating the stress field as a function of the crack propagation by using the Fracture Analysis Code (Franc3D®, http://www.fracanalysis.com/Fracture Analysis Consultants, Inc 121 Eastern Heights Dr Ithaca, 14850 New York, NY, USA. Phone: 607-257-4970).

Author Contributions: The three co-guest-editors of this special issue shared the editorial duties, managing the review process for the papers considered for publication. All authors have read and agreed to the published version of the manuscript.

Funding: This research received no external funding.

Acknowledgments: The editors would like to express their thanks to all authors of the Special Issue for their valuable contributions and to all reviewers for their useful efforts to provide valuable reviews. We expect that this Special Issue offers a timely view of advanced topics in the structural behaviour of advanced materials, which will stimulate further novel academic research and innovative applications.

Conflicts of Interest: The authors declare no conflict of interest.

References

1. Parvez, M.; Chen, Y.; Karnati, S.; Coward, C.; Newkirk, J.; Liou, F. A Displacement Controlled Fatigue Test Method for Additively Manufactured Materials. *Appl. Sci.* **2019**, *9*, 3226. [CrossRef]

2. Walker, P.; Malz, S.; Trudel, E.; Nosir, S.; ElSayed, M.; Kok, L. Effects of Ultrasonic Impact Treatment on the Stress-Controlled Fatigue Performance of Additively Manufactured DMLS Ti-6Al-4V Alloy. *Appl. Sci.* **2019**, *9*, 4787. [CrossRef]

3. Yue, C.; Gao, H.; Liu, X.; Liang, S. Part Functionality Alterations Induced by Changes of Surface Integrity in Metal Milling Process: A Review. *Appl. Sci.* **2018**, *8*, 2550. [CrossRef]

4. Lan, T.; Pan, C.; Liu, P.; Chou, M. Fracture Resistance of Monolithic Zirconia Crowns on Four Occlusal Convergent Abutments in Implant Prosthesis. *Appl. Sci.* **2019**, *9*, 2585. [CrossRef]

5. Akama, M. Fatigue Crack Growth under Non-Proportional Mixed Mode Loading in Rail and Wheel Steel Part 1: Sequential Mode I and Mode II Loading. *Appl. Sci.* **2019**, *9*, 2006. [CrossRef]

6. Akama, M.; Kiuchi, A. Fatigue Crack Growth under Non-Proportional Mixed Mode Loading in Rail and Wheel Steel Part 2: Sequential Mode I and Mode III Loading. *Appl. Sci.* **2019**, *9*, 2866. [CrossRef]

7. Canale, G.; Kinawy, M.; Maligno, A.; Sathujoda, P.; Citarella, R. Study of Mixed-Mode Cracking of Dovetail Root of an Aero-Engine Blade Like Structure. *Appl. Sci.* **2019**, *9*, 3825. [CrossRef]

Article

A Displacement Controlled Fatigue Test Method for Additively Manufactured Materials

Mohammad Masud Parvez [1,*], Yitao Chen [1], Sreekar Karnati [1], Connor Coward [1], Joseph W. Newkirk [2] and Frank Liou [1]

1   Department of Mechanical and Aerospace Engineering, Missouri University of Science and Technology, Rolla, MO 65401, USA
2   Material Science and Engineering, Missouri University of Science and Technology, Rolla, MO 65401, USA
*   Correspondence: mphf2@umsystem.edu; Tel.: +1-573-202-1506

Received: 6 July 2019; Accepted: 31 July 2019; Published: 7 August 2019

Abstract: A novel adaptive displacement-controlled test setup was developed for fatigue testing on mini specimens. In property characterization of additive manufacturing materials, mini specimens are preferred due to the specimen preparation, and manufacturing cost but mini specimens demonstrate higher fatigue strength than standard specimens due to the lower probability of material defects resulting in fatigue. In this study, a dual gauge section Krouse type mini specimen was designed to conduct fatigue tests on additively manufactured materials. The large surface area of the specimen with a constant stress distribution and increased control volume as the gauge section may capture all different types of surface and microstructural defects of the material. A fully reversed bending (R = −1) fatigue test was performed on simply supported specimens. In the displacement-controlled mechanism, the variation in the control signal during the test due to the stiffness variation of the specimen provides a unique insight into identifying the nucleation and propagation phase. The fatigue performance of the wrought 304 and additively manufactured 304L stainless steel was compared applying a control signal monitoring (CSM) method. The test results and analyses validate the design of the specimen and the effective implementation of the test bench in fatigue testing of additively manufactured materials.

Keywords: adaptive control; fatigue testing; simply supported bending; mini specimen; additive manufacturing; 304L stainless steel

---

1. Introduction

Fatigue is a progressive and permanent structural change due to fluctuating stresses or strains subjected to a material. 50% to 90% of mechanical failures of structures are due to fatigue [1,2]. Fatigue test is indispensable in the characterization of materials but the test is both time-consuming and very expensive [3,4]. In this research, a unique test setup was designed and developed to reduce the test cost using mini specimen. The measured strength of a material subjected to monotonic or cyclic loading depends inversely on the specimen size. The impact of the size effect on mechanical properties depends on the type and local feature of the material structure i.e., grain size, microcracks, inclusions, discontinuities, dislocations, and other defects [5–7]. Extended studies were carried out to investigate the effect of specimen size and loading condition on fatigue behavior of metallic materials [8–15]. Statistically, large specimens contain more extreme defects. The presence of larger defects leads to crack growth and failure at lower stress levels. Sun [16] proposed a probabilistic method to correlate the effects of specimen geometry and loading condition on the fatigue strength based on the Weibull distribution. Tomaszewski [4] performed comparative tests on mini specimens and normative specimens, and verified the monofractal approach based on Basquin's equation along with the Weibull weakest link model. There are some other statistical methods proposed to evaluate the size effect on the fatigue test [17–21]. All of these approaches epitomize that standard specimens demonstrate lower fatigue strength than mini specimens due to the higher probability of larger material defects. Additively manufactured materials have a higher probability of defects compared to wrought materials. In this paper, the implementation of a dual gauge section Krouse type mini specimen increases the surface area to capture all different types of surface and microstructural defects since most of the fatigue failures are initiated at the surface or subsurface due to the presence of defects.

Geometrically, the size effect is related to the nonlinear distribution of the stress [22–24]. The stress gradient occurring under bending and shear stress has a higher influence on the size effect for a bending type test compared to axially loaded cyclic test but the axial fatigue test on mini specimens suffers buckling. In this study, the transverse bending test with a constant stress distribution within the gauge section in a specimen eliminates the stress gradient effect.

There are several techniques already developed to monitor the crack nucleation and propagation during the fatigue test. These techniques include the acoustic emission diagnostic method [25–27], electrical resistance change method [28,29], meandering winding magnetometer (MWM)-array eddy current sensing [30], and thermographic method [31]. All of these techniques require an additional sensor with intensive signal processing. In the current work, we introduce a simple but effective control signal monitoring (CSM) method to identify the nucleation and propagation phase. In a displacement-controlled mechanism, the control signal decreases with the decrease in the structural stiffness of the specimen. The change in the control signal provides insight in estimating the nucleation and propagation phase. In this study, the fatigue test was conducted on wrought 304 and additively manufactured 304L stainless steel specimens. The CSM method was applied to identify the nucleation and propagation phase. The test results were compared to validate the design of the specimen and the test setup performance in high cycle fatigue testing.

2. Methodology

In this study, a fully reversed bending (R = −1) fatigue test was performed on simply supported specimens. A simply supported testing methodology has several advantages over a fully clamped type of loading mechanism. The maximum deflection in a simply supported and a fully clamped beam with a concentrated load F at the center are given by Equations (1) and (2) respectively [32,33],

$$\delta_{max} = \frac{Fl^3}{48EI} \tag{1}$$

$$\delta_{max} = \frac{Fl^3}{192EI} \tag{2}$$

where F, δ_{max}, l, E, and I are the applied force, maximum deflection, length, modulus of elasticity, and moment of inertia of the beam respectively. For a given load, the displacement is four times higher in a simply supported bending than in a fully clamped bending. During the fatigue test, investigators attempt to actuate the specimen at its natural frequency to achieve maximum displacement. However, the dynamics of the actuator coupled with the specimen limit the operation. Therefore, as an alternate, we adopted a simply supported bending mechanism as the testing methodology.

3. Specimen Design, Analysis and Preparation

3.1. Design of the Specimen

A dual gauge section Krouse type mini specimen was designed for simply supported loading. The specimen is a modified form of the ASTM (American Society for Testing and Materials) International standard B593-96(2014)e1, definition E206, and practice E468 [34]. Some authors already reported on the modification and implementation of the specimen in miniature form [35–38]. Since the specimens are miniature size, Haydirah [39] performed an error analysis based on the effect of specimen's dimension. Figure 1 shows the dimension of our specimen. The effective length between both clamping ends is 25.4 mm. Each gauge is 4.34 mm long. The total gauge covers 34.17% of the total effective length of the specimen. The dual gauge increases the overall surface area. The failure is expected to be within the gauges. Another reason for choosing the dual gauge is to maintain symmetry. In a single cantilever beam, the actuator follows a curved path during excitation. To keep the path of the actuator one dimensional, and to distribute the load symmetrically along with the specimen, the dual gauge concept is opted.

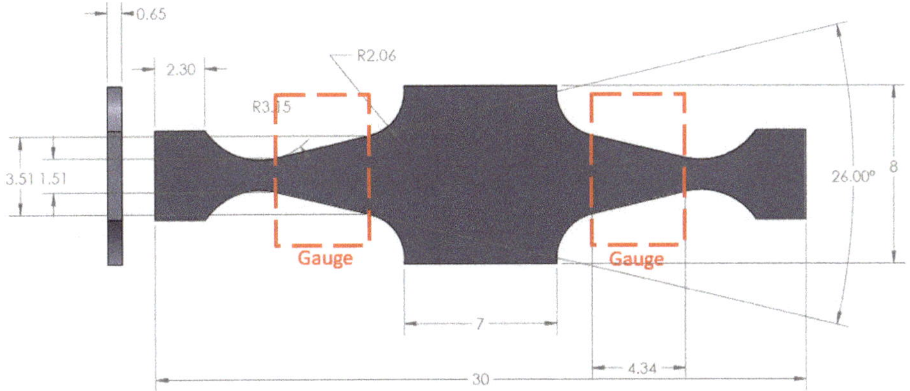

Figure 1. Drawing of the dual gauge section Krouse type mini specimen, all units are in mm.

3.2. Stress Calculation

Previous studies showed that simple beam equation is applicable to calculate the stress in miniature wedge shaped specimen [34–39]. The stress in a simply supported bending beam with a point load at the center can be expressed as [40],

$$\sigma = \frac{M(x)}{I(x)}\frac{h}{2} \tag{3}$$

where, σ, $M(x)$, $I(x)$, and h are the stress, moment, second moment of inertia, and the thickness of the specimen, respectively. For a simply supported beam, $M(x) = \frac{Fx}{2}$, and $I(x) = \frac{b(x)h^3}{12}$ where, F, and b are the point load, and the width of the specimen, respectively. For a Krouse type specimen $b(x) = 2kx$ where, k is the slope of the specimen. Inserting $M(x)$, and $I(x)$ in Equation (3), we get,

$$\sigma = \frac{3F}{2kh^2} = j(F, h) \tag{4}$$

where, j is the stress function. The nominal stress σ within the gauge in Equation (4) depends on the force applied and the thickness of the specimen, not on the distance x. Ideally, a constant stress distribution is expected but in reality at the defect zone or at the lower strength site, the actual local stress will be higher than the nominal stress.

3.3. Sensitivity and Uncertainty Analysis

The specimen is a miniature size compared to the standard one. The necessity of sensitivity and uncertainty analysis is inevitable to determine the optimal thickness of the specimen. The stress calculation is sensitive to the force and thickness of the specimen according to Equation (4). Uncertainty in force measurement depends on the sensor's accuracy, calibration, and set up. The thickness is sensitive to the machining and polishing process. For a higher thickness, a higher force is required to attain particular stress. This leads to the necessity of a high power system and actuator. An optimal thickness was determined to eliminate the necessity of high power fatigue machine and external cooling. Partially differentiating Equation (4) we get,

$$\nabla \bar{j} = \begin{bmatrix} \frac{F}{\sigma} \times \frac{\partial \sigma}{\partial F} \\ \frac{h}{\sigma} \times \frac{\partial \sigma}{\partial h} \end{bmatrix} = \begin{bmatrix} 1 \\ -2 \end{bmatrix} \tag{5}$$

From Equation (5), we can see 1% variation in specimen thickness produces 2% change in stress value. To estimate the thickness uncertainties, 10 specimens were prepared. The thickness was measured using a high precision laser displacement sensor. The uncertainty was calculated obtaining overall standard deviation (std) using Equation (6).

$$std = \frac{1}{n} \sum_{j=1}^{n} (x_j - \bar{x})^2 = \frac{1}{n} \left[\sum_{i=1}^{g} n_i S_i^2 + \sum_{i=1}^{g} (\bar{x}_i - x)^2 \right] \; textrmwhere, \bar{x} = \frac{\sum_{i=1}^{g} n_i x_i}{n} \tag{6}$$

where, $\bar{x}_i$, S_i, and n_i are the mean, standard deviation, and the number of scanned data points of i th specimen respectively. $\bar{x}$ is the overall mean, and n is the total number of data points. For $\pm 5\%$ stress variation, the calculated optimal thickness of the specimen was 0.509 mm with three sigma quality level. Including a factor of safety, the specimen thickness used in this study is 0.65 mm.

3.4. Finite Element Analysis

Finite element analysis was performed using ABAQUS 2018 software (Dassault Systèmes Simulia Corp; Providence, RI, USA) to demonstrate the constant stress distribution within gauge sections. According to the specimen design, as shown in Figure 2, the 3D prototype of the specimen was simply supported at both sides which are marked by red lines ($U_z = 0$). To ensure a symmetric deformation, the displacement on center-lines along the x-axis (green line) and y-axis (blue line) are restricted in y direction ($U_y = 0$) and x direction ($U_x = 0$), respectively. A constant displacement $U_z = 0.150$ mm was applied on the 3 mm $\times$ 7 mm dark grey rectangular area at the center of the specimen, which indicates the rectangular plate washer in the machine setup. Boundary conditions are listed in the box under the 3D prototype of the specimen. The Young's modulus and Poisson's ratio set for the wrought 304 stainless steel were 200 GPa and 0.3 respectively. A linear elastic model was applied to observe the mechanical response under this static condition, as the deformation is within the elastic regime. The distribution of the nominal stress S11 on the whole specimen is then obtained by the simulation, as shown in Figure 3. A constant nominal stress within triangular gauge sections can be observed, and it reaches the maximum value at the surface. Convergence study was also performed by selecting 6 different mesh sizes which result in the number of elements ranging from 1263 to 166,506. The data points in Figure 4 shows that the nominal stress converges to approximately 177.7 MPa as the number of elements increases to 166,506, since when the number of mesh elements increases from 76,698 to 166,506, the change in nominal stress value is less than 0.2%. Figure 3 exhibits the nominal stress distribution with the number of elements of 166,506. The sole purpose of using FEA analysis is to demonstrate the stress distribution within the gauge.

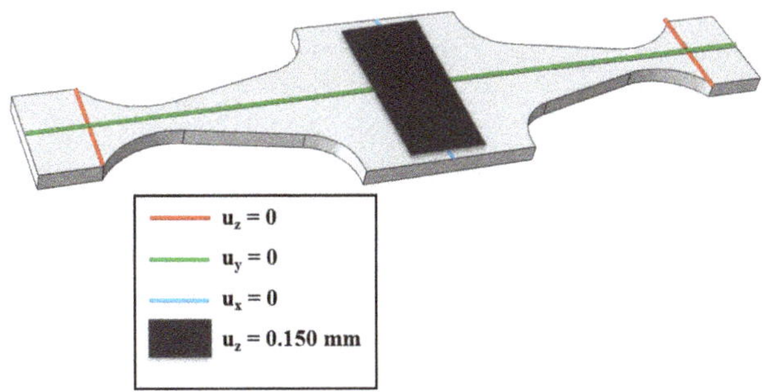

Figure 2. FEA simulation setup for wrought 304 stainless steel specimen.

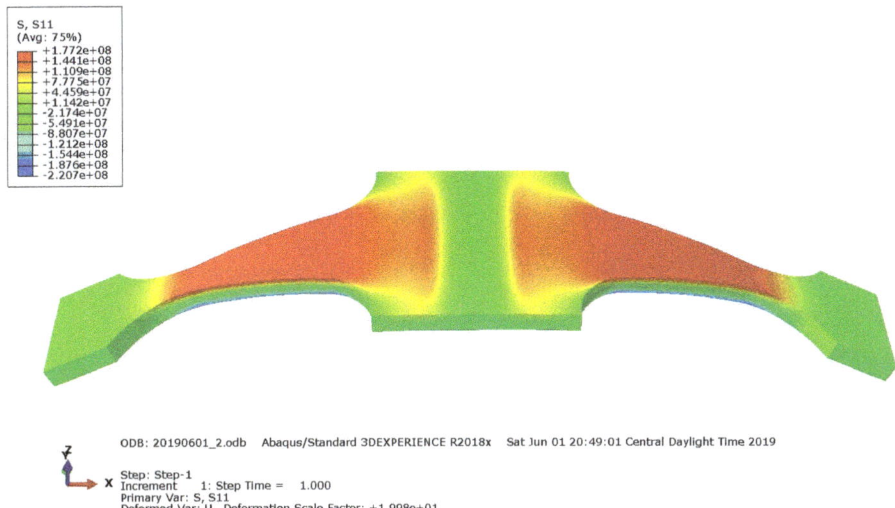

Figure 3. FEA simulation result of the specimen. The red zone on the top surface shows the stress distribution is constant within the gauges.

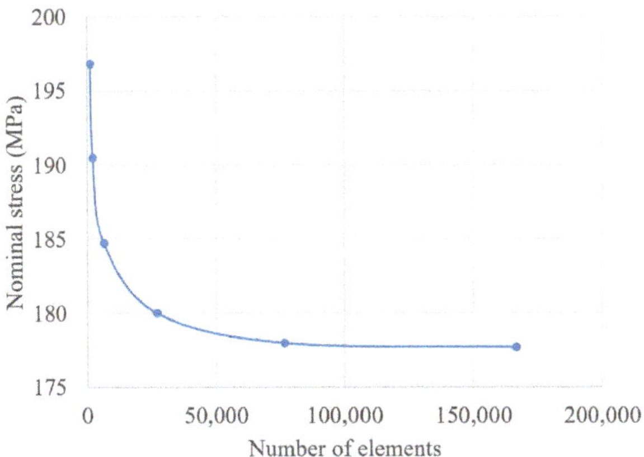

Figure 4. Convergence analysis of the FEA simulation results.

3.5. Materials and Specimen Preparation

The materials tested in this study are hot rolled and annealed 304 stainless steel bulk material and additive manufacturing (AM) fabricated 304L SS bar. These materials were chosen because they are economical and widely used due to their strength and high resistance to corrosion. The chemistry of both the wrought material and powder used as the feedstock for AM is listed in Table 1. The relatively close chemistry of both materials except Ni which is 2% higher but not expected to make a significant difference in the test results may aid a better understanding of the comparative study. Rough finish (average $R_a = 3.82$ µm) and fine finish (average $R_a = 0.482$ µm, average $R_z = 4.242$ µm) wrought specimens were machined using W-EDM, while additively manufactured fine finish specimens were cut along Z axis from a bar fabricated using the selective laser melting (SLM) process. A Renishaw AM250 machine was used to build the part. An optimal process parameter listed in Table 2 was applied to yield maximum part density. A total of 10 specimens for each type were manufactured with no additional surface preparation.

Table 1. Chemical properties (wt%) of 304L stainless steel powder and bulk 304 stainless steel.

Material	C	Mn	Si	S	P	Cr	Ni	Cu	Mo	Co	N	O
Wrought	0.023	1.69	0.43	0.020	0.034	18.10	8.02	0.63	0.24	0.15	0.084	-
Powder	0.015	1.40	0.63	0.004	0.012	18.50	9.90	<0.1	-	-	0.090	0.02

Table 2. Parameters used to build additively manufactured part using selective laser melting (SLM) process.

Parameter Set	Power (watt)	Hatch Space (µm)	Point Distance (µm)	Exposure Time (millisecond)	Energy Density (MJ/m^3)	Raster Rotation (degree)
Nominal	200	85	60	75	58.8	67

4. Experimental Setup

4.1. Mini Fatigue Testing Machine

The mini fatigue testing machine consists of six major parts: (i) an electromagnetic actuator, (ii) a non-contact displacement sensor, (iii) a load cell, (iv) a controller, (v) a power amplifier or driver, and (vi) a test bench. The voice coil of a subwoofer was used as the actuator. The sub-woofer behaves as a low audio frequency shaker. The mathematical model of an electrodynamic shaker and a sub-woofer is relatively similar though the moving elements of a shaker are more rigid than a subwoofer. Higher rigidity multiplies the power requirements. To design a low power system, a soft mechanical suspension of the sub-woofer was implemented to transfer maximum energy to the specimen. The actuator is made of a high Curie temperature ferrite magnet with a cast aluminum frame of 10 inches diameter. The larger diameter of the voice coil (3 inches) than the length (1 inch) of the specimen supports the one-dimensional movement. The dust cap of the voice coil was replaced with a plastic flange. A load cell was mounted in-line between the central clamp and the flange to measure the tensile and compressive force. To measure the displacement of the specimen, a high-speed non-contact laser displacement sensor was fixed with a guide rail. Figure 5 illustrates the test bench setup. First, the specimen was clamped at the center. The specimen sits on the bearings at both ends as shown in Figure 5. Spacers were used at both ends to ensure no preload on the specimen. Then, the other bearing holders were placed and clamped using heavy load toggle clamps on top. To ensure line contact at both ends, bearings were used. Bearings also minimize the friction during the simply supported vibration test. By sliding the displacement sensor using the guide rail, the sensor was pointed at the center of the specimen. The displacement measured by the sensor was processed using a microcontroller to determine the amplitude and mean of the displacement. The data was sent to a computer from the microcontroller using a serial port. An adaptive controller was implemented in the Python development environment to estimate the required amplitude of the sinusoidal control signal. The signal from the computer was sent to a waveform generator via Ethernet. A linear power amplifier connected with the waveform generator drives the actuator. All process and manipulated variables were stored for further analysis to identify the nucleation and propagation phase.

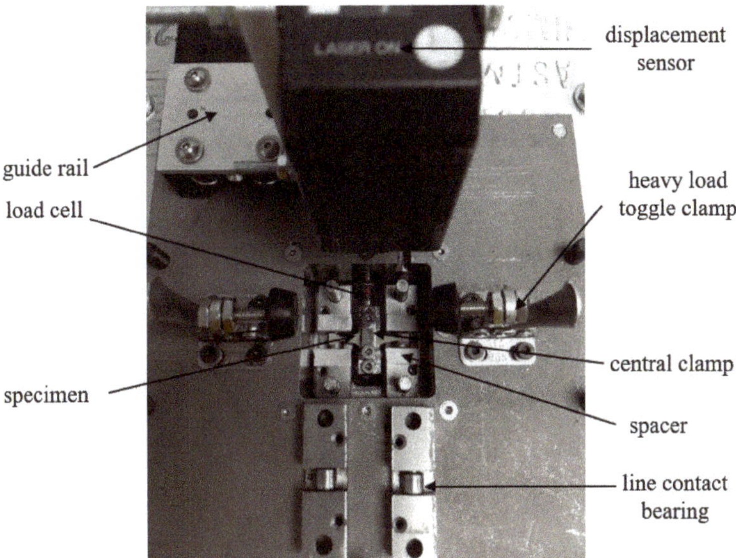

Figure 5. Fatigue test bench with a specimen mounted.

4.2. Adaptive Controller Design

An adaptive proportional and derivative (PD) controller was designed to control the displacement amplitude. A conventional PID controller was avoided since the system parameters change due to the structural stiffness change of the specimen during the test. Material hardening or softening may occur too. There may also be a possibility of overshoot above the set point during the transient condition. Overshoot may affect the test results. Therefore, an adaptive controller was designed. The design of an adaptive controller follows Equation (7).

$$u(k) = u(k-1) + P * error + D * \frac{error(k) - error(k-1)}{\Delta t} \tag{7}$$

where, $u(k)$, and $u(k-1)$ are the control signal amplitudes at time k, and $(k-1)$ respectively. P, and D are the proportional and derivative gain respectively, and Δt is the time step. The proportional controller offsets the current value linearly with the error, and the derivative controller adds in controlling the actuation based on the rate of the change of error. The error is defined as,

$$error = d^{set}_{p-p} - d^{current}_{p-p} \tag{8}$$

where, d^{set}_{p-p} and $d^{current}_{p-p}$ are the desired and current displacement amplitude respectively. The controller values were chosen by manual tuning with caution that no overshoot occurs above the set point. The controller values need to be varied with the test frequency and test material as well. The P and D controllers were set at 5.0 and 0.5 respectively for 304 materials at 56 Hz test frequency.

5. Results and Discussion

The closed-loop displacement-controlled fatigue test was performed on wrought 304 and SLM fabricated 304L SS specimens. The sinusoidal excitation frequency was set at 56 Hz. Figure 6 shows the displacement amplitude of a specimen actuated at 0.100 volts control signal amplitude for different frequencies. The system response is a window function with 39 Hz cutoff frequency. In terms of system dynamics, the fatigue test is a harmonic forced vibration of two mass-spring systems. One is the actuator, and another is the specimen. At cutoff frequencies, the system behaves like a shock absorber. A detailed explanation can be found in the literature [41]. The frequency response shows that the displacement amplitude is maximum at 56 Hz, 95 Hz, and 134 Hz. The test frequency was chosen 56 Hz since external cooling may be required at higher frequencies. All experiments were conducted for simply supported fully reversed bending test at room temperature. During the test, the temperature of the specimen was monitored using an infrared temperature sensor. The deviation in the temperature remains within $\pm 2\,^\circ$C at 56 Hz test frequency.

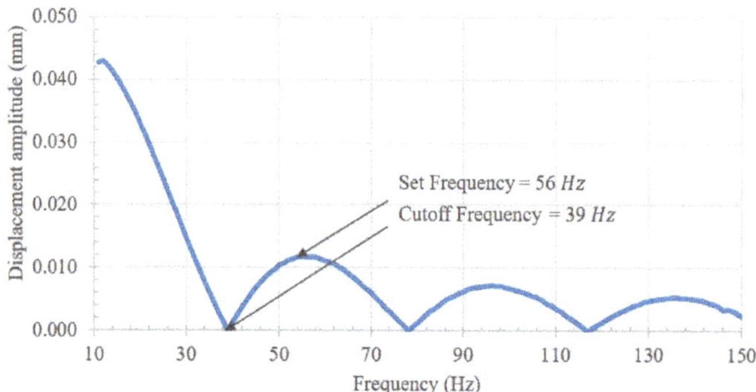

Figure 6. The frequency response of a specimen actuated at 0.100 volts control signal amplitude.

In a Krouse type specimen, the fatigue failure can occur at any location within the gauge. All the specimens tested in this study failed within the gauge as expected. The random failure location is due to the defects present randomly within the gauge. The nominal stress distribution is supposed to be constant while the local stress is expected to be high at the defect zone. Figure 7 exhibits the failure location of wrought specimens. Figure 8 illustrates the displacement and control signal amplitude for the wrought specimen actuated at 0.200 mm amplitude which corresponds to 514.26 MPa nominal stress. During the test, the stiffness of the specimen decreases as the crack grows, propagates, and final failure occurs. The control signal amplitude decreases with the reduction of stiffness to maintain the desired set displacement. The displacement amplitude increases suddenly during the final failure. The test was stopped automatically when the amplitude was above a threshold. The test result at different displacements illustrated in Figure 9 validates the effective performance of the adaptive controller.

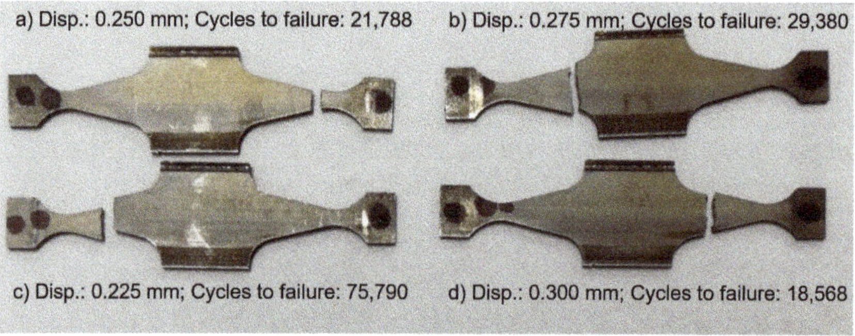

Figure 7. Fatigue failure of specimens actuated at different displacement amplitude.

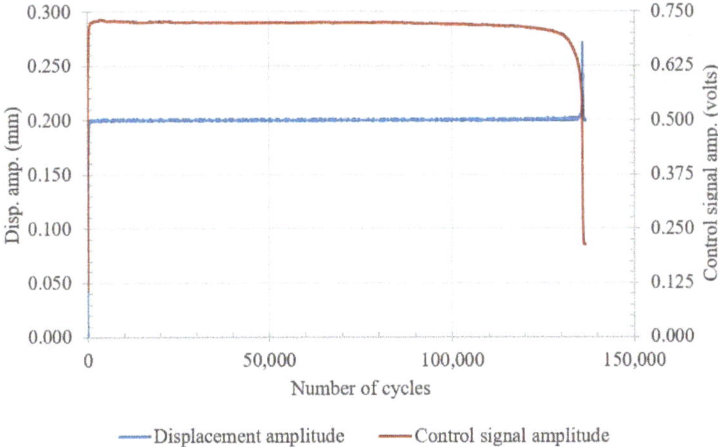

Figure 8. Displacement and control signal amplitude up to the entire fatigue life cycle of a fine finished wrought specimen.

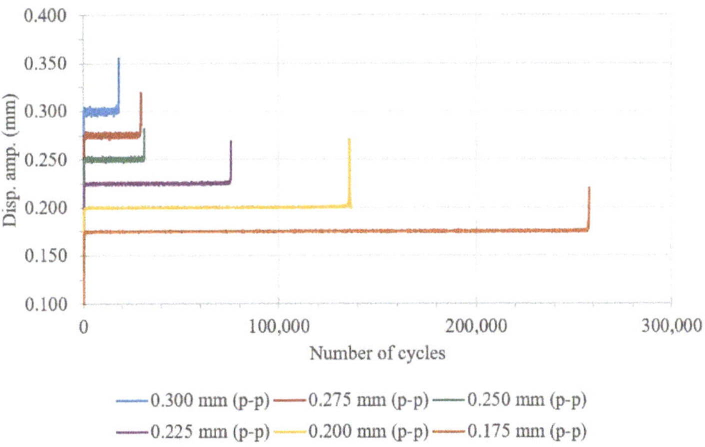

Figure 9. Displacement amplitude control of fine finished wrought specimens up to the entire fatigue life cycle.

In this study, we introduce a control signal monitoring (CSM) method to identify the nucleation and propagation stage. During the fatigue test, the crack first grows slowly, which is termed as nucleation. When the nucleation process ends, the crack starts propagating and the final failure occurs. The stiffness of the specimen decreases at all stage, but the rate of the stiffness change is different. Wang [42] reported for carbon fiber polymer-matrix composite that the stiffness of the material has an inverse analogous relation with the change in resistance up to the end of nucleation phase while performing the electrical potential technique on fatigue test to identify the nucleation and propagation phase. Grammatikos [31] implemented the linear regression analysis on the relative potential change as a function of fatigue life fraction to identify different stages. Similarly, this is possible to identify the phases using linear regression analysis on the control signal. First linear regression was applied on the control signal. Then the peak amplitude of the signal near the end of regression line was marked as the end of nucleation phase, as shown in Figure 10. Here the regression helps in choosing the peak of the signal. Current study demonstrates the implementation of CSM method. Future study includes the sensitivity analysis on the monitoring signal. The implementation of the CSM method on other wrought and SLM specimens is shown in Figure 11. To determine the maximum nominal stress, the average of the peak load was calculated up to the end of the nucleation phase. Inserting the average in Equation (4), the nominal stress was calculated. Figure 12 shows the tensile and compressive load response up to the final failure for the wrought specimen displaced at 0.200 mm amplitude.

The fatigue life of a specimen in terms of the number of cycles is the sum of cycles during nucleation and propagation stages. Materials demonstrate a higher life cycle at low-stress value. Figure 13 shows the end of the nucleation phase cycle and cycles to failure for fine finished wrought and SLM specimens. The trend of nucleation and propagation presented in the literature [43,44] supports the results. Both the wrought and SLM specimens demonstrate an increase in the nucleation and propagation cycle as the stress goes low, but for a particular stress value, both the nucleation and propagation cycles for the SLM material is lower than those for the wrought material. A possible reason is that additively manufactured materials have a higher probability of different types of defects. The fracture surface analysis of wrought and SLM specimens is shown in Figure 14. In the SLM specimen, the crack is initiated at the lack of fusion defect near the top surface while surface defect is the crack initiation source for wrought specimens. We observed similar type of defects to be the crack nucleation site for other SLM specimens. The crack starts growing earlier in SLM materials than in wrought materials. Therefore, the nucleation life cycle is less. The presence of other defects such as micro-cracks, pores, and lack of fusion within the volume enhance the propagation rate. Additionally, some authors reported that inter-layer bonding is weak in SLM materials [45,46]. Therefore, SLM fabricated 304 L SS demonstrate lower fatigue strength than bulk material.

Further analysis was performed to evaluate the CSM method in identifying nucleation and propagation phase. Rough finished and notched rough finished wrought specimens were prepared to conduct fatigue test. Rough finished specimens were chosen because these are easy to prepare. Figure 15 shows the notch location in the specimen. Figure 16 illustrates the nucleation and propagation phase identified using the CSM method for the specimens. As we know, notched specimens have higher stress concentration, hence, they fail earlier. The number of cycles decreases more in nucleation phase due to the notch while the influence of notch on the propagation is minimal. This attributes to the proper implementation of the CSM method in identifying the end of nucleation phase. In future, the method will be validated determining the crack length at the end of nucleation phase.

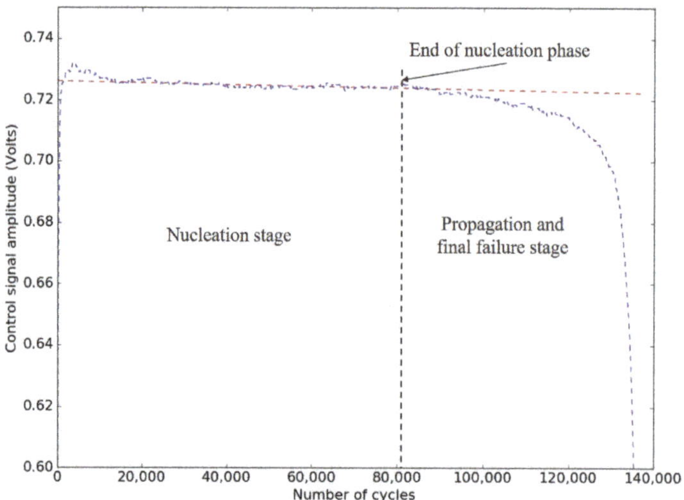

Figure 10. Nucleation and propagation stage of a fine finished wrought specimen displaced at 0.200 mm amplitude.

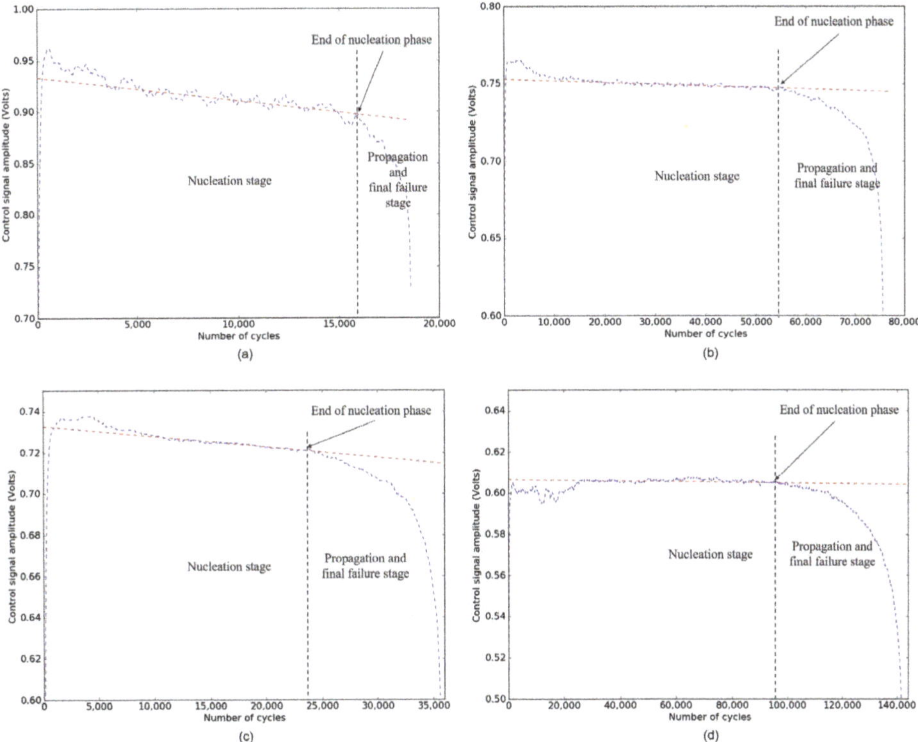

Figure 11. Nucleation and propagation stage of fine finished specimens, (**a**) wrought specimen displaced at 0.300 mm amplitude; (**b**) wrought specimen displaced at 0.225 mm amplitude; (**c**) SLM specimen displaced at 0.200 mm; (**d**) SLM specimen displaced at 0.175 mm amplitude.

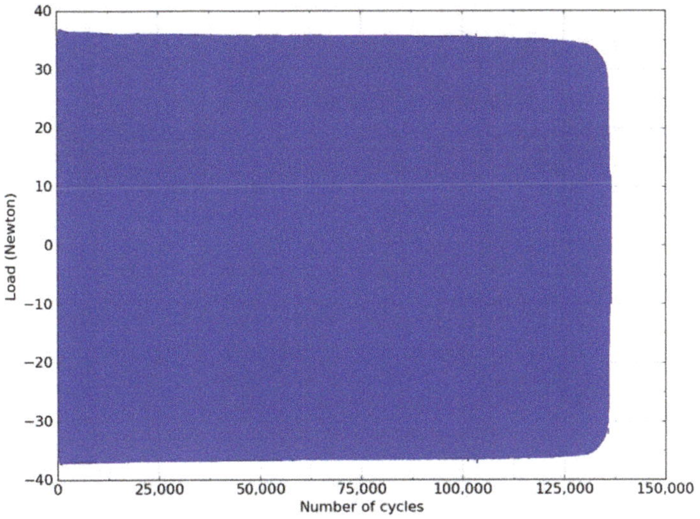

Figure 12. Load values for the specimen displaced at 0.200 mm amplitude.

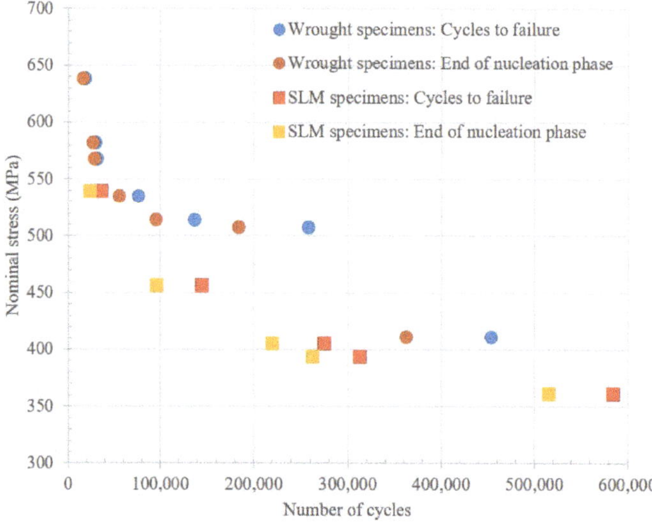

Figure 13. End of nucleation phase and cycles to failure for fine finished wrought and SLM fabricated specimens.

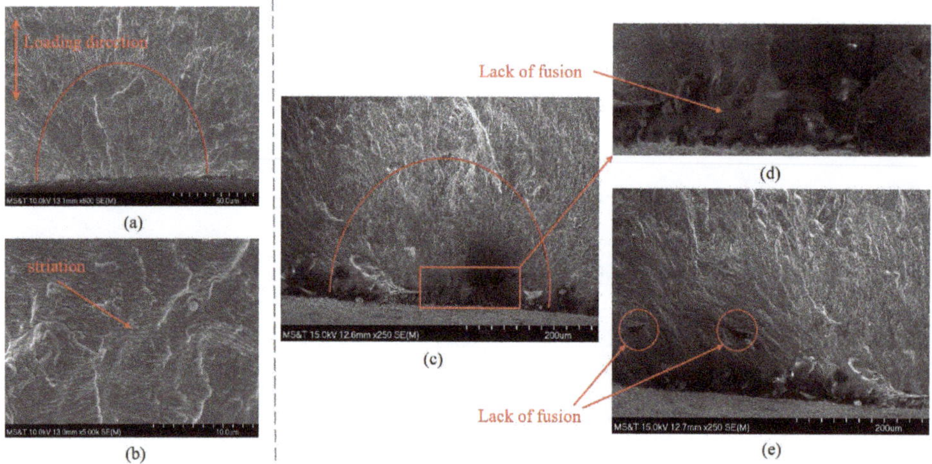

Figure 14. Fracture surface analysis of fine finished wrought and SLM fabricated materials, (**a**) crack nucleation site; and (**b**) crack propagation site of a wrought specimen, the nominal stress is 411.67 MPa; (**c**) crack nucleation site of SLM fabricated specimen, the nominal stress is 394.76 MPa; (**d**) presence of lack of fusion at the surface initiating the crack; (**e**) presence of lack of fusion within the volume enhancing crack propagation rate.

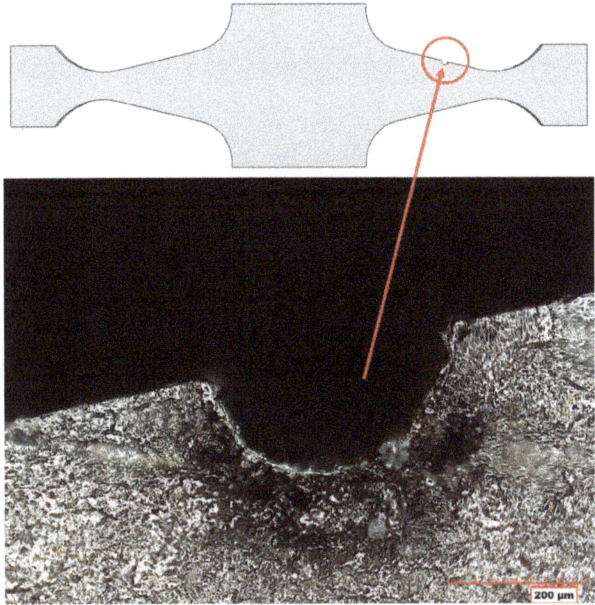

Figure 15. Notched rough finished wrought specimens prepared with a W-EDM wire radius of 0.125 mm. The average radius of the notch measured was 0.180 mm.

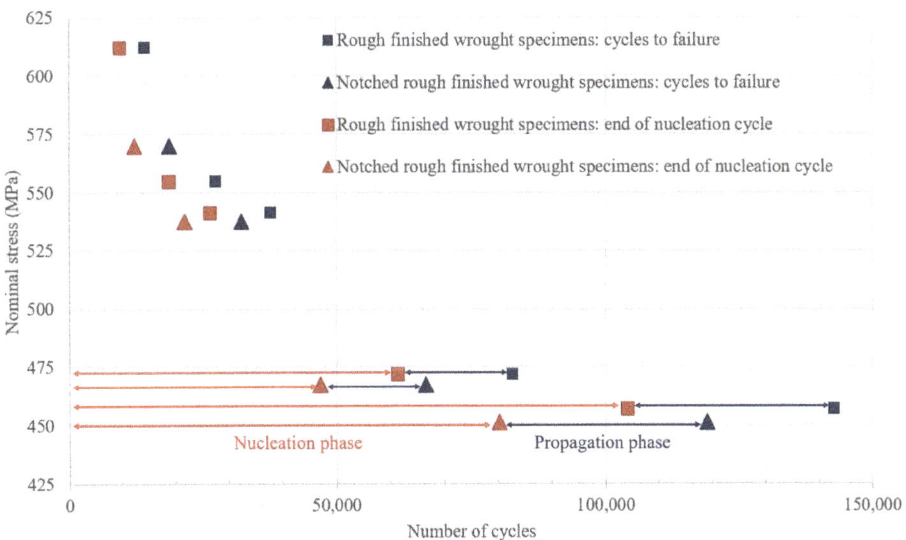

Figure 16. Nucleation and propagation phase of rough finished wrought and notched rough finished wrought specimens.

The Wohler curve for fine finished and rough finished wrought specimens, and fine finished SLM specimens was plotted as shown in Figure 17. Both fine finished SLM and rough finished wrought materials exhibit low fatigue strength compared to the fine finished wrought material. The endurance limit (10^7) of the fine finished, rough finished wrought specimen and fine finished SLM specimen reported here are 404.26, 344.62 and 336.52 MPa respectively. The yield tensile strength (YTS) and ultimate tensile strength (UTS) for the wrought material used in this study are 582.5 and 780.9 respectively while for the SLM material with the same process parameter used, they are 368.4 and 536.7 MPa respectively [47]. These results are in good agreement with the general relationship between fatigue limit and ultimate tensile strength [43]. The endurance limit is also comparable with the results reported by other authors [38,48].

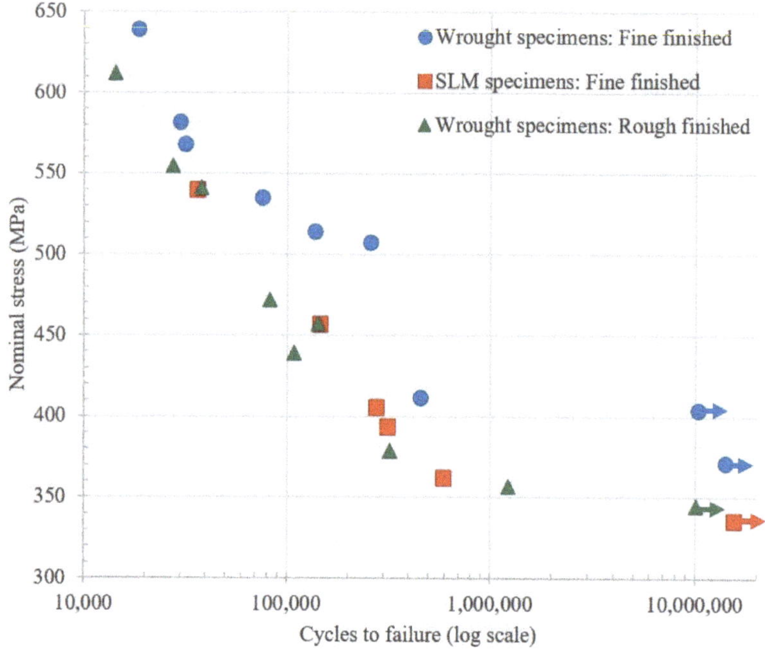

Figure 17. Wohler curve plot of wrought and SLM fabricated specimens.

6. Conclusions

In this study, a dual gauge section Krouse type mini specimen was designed to achieve a constant stress distribution with increased volume to conduct fatigue test. The test was performed with a simply supported loading mechanism on wrought and additively manufactured materials using a unique adaptive displacement controlled mini fatigue test set up. A new diagnosis method named control signal monitoring (CSM) was employed to identify the nucleation and propagation stages. The test results and analyses illustrate that SLM fabricated 304L stainless steel demonstrate lower fatigue strength in terms of both the nucleation and propagation cycles compared to bulk wrought material. The test method developed here can be applied in the extensive study on other additively manufactured materials in the future.

Author Contributions: The contribution of each author is summarized as conceptualization, M.M.P. and C.C.; methodology, M.M.P. and S.K.; software, M.M.P.; data curation, M.M.P.; validation M.M.P.; formal analysis, M.M.P. and Y.C.; investigation, M.M.P.; writing—original draft, M.M.P.; writing—review and editing, M.M.P., Y.C., C.C., J.W.N., F.L.; visualization, M.M.P.; supervision, J.W.N. and F.L.; project administration, F.L.; funding acquisition, F.L.

Funding: This research was supported by National Science Foundation Grant CMMI-1625736. Part of the work was also funded by the Department of Energy's Kansas City National Security Campus which is operated and managed by Honeywell Federal Manufacturing Technologies, LLC under contract number DE-NA0002839.

Conflicts of Interest: The authors declare no conflict of interest.

References

1. Santecchia, E.; Hamouda, A.; Musharavati, F.; Zalnezhad, E.; Cabibbo, M.; El Mehtedi, M.; Spigarelli, S. A review on fatigue life prediction methods for metals. *Adv. Mater. Sci. Eng.* **2016**, *2016*, 9573524. [CrossRef]
2. Campbell, F.C. *Elements of Metallurgy and Engineering Alloys*; ASM International: Cleveland, OH, USA, 2008; Chapter 14; p. 245.
3. Sempruch, J.; Tomaszewski, T. Application of mini specimens to high-cycle fatigue tests. *J. Pol. CIMAC* **2011**, *6*, 279–287.
4. Tomaszewski, T.; Sempruch, J. Verification of the fatigue test method applied with the use of mini specimen. In *Key Engineering Materials*; Trans Tech Publications: Zurich, Switzerland, 2014; Volume 598, pp. 243–248.
5. Bažant, Z.P. Size effect in blunt fracture: Concrete, rock, metal. *J. Eng. Mech.* **1984**, *110*, 518–535. [CrossRef]
6. Hirose, T.; Sakasegawa, H.; Kohyama, A.; Katoh, Y.; Tanigawa, H. Effect of specimen size on fatigue properties of reduced activation ferritic/martensitic steels. *J. Nucl. Mater.* **2000**, *283*, 1018–1022. [CrossRef]
7. Haftirman, H. The size effect on fatigue strenght of structural steel materials in high-humidity environment. *J. Mek.* **2011**, *32*, 1–11.
8. Beretta, S.; Ghidini, A.; Lombardo, F. Fracture mechanics and scale effects in the fatigue of railway axles. *Eng. Fract. Mech.* **2005**, *72*, 195–208. [CrossRef]
9. Nakajima, M.; Tokaji, K.; Itoga, H.; Shimizu, T. Effect of loading condition on very high cycle fatigue behavior in a high strength steel. *Int. J. Fatigue* **2010**, *32*, 475–480. [CrossRef]
10. Akiniwa, Y.; Miyamoto, N.; Tsuru, H.; Tanaka, K. Notch effect on fatigue strength reduction of bearing steel in the very high cycle regime. *Int. J. Fatigue* **2006**, *28*, 1555–1565. [CrossRef]
11. Wormsen, A.; Sjödin, B.; Härkegård, G.; Fjeldstad, A. Non-local stress approach for fatigue assessment based on weakest-link theory and statistics of extremes. *Fatigue Fract. Eng. Mater. Struct.* **2007**, *30*, 1214–1227. [CrossRef]
12. Findley, W. An explanation of size effect in fatigue of metals. *J. Mech. Eng. Sci.* **1972**, *14*, 424–428. [CrossRef]
13. Zhu, S.P.; Foletti, S.; Beretta, S. Evaluation of size effect on strain-controlled fatigue behavior of a quench and tempered rotor steel: Experimental and numerical study. *Mater. Sci. Eng. A* **2018**, *735*, 423–435. [CrossRef]
14. Diemar, A.; Thumser, R.; Bergmann, J. Determination of local characteristics for the application of the Weakest-Link Model. *Materialwissenschaft Werkstofftechnik Entwicklung Fertigung Prüfung Eigenschaften Anwendungen Technischer Werkstoffe* **2005**, *36*, 204–210. [CrossRef]
15. Leitner, M.; Garb, C.; Remes, H.; Stoschka, M. Microporosity and statistical size effect on the fatigue strength of cast aluminium alloys EN AC-45500 and 46200. *Mater. Sci. Eng. A* **2017**, *707*, 567–575. [CrossRef]
16. Sun, C.; Song, Q. A method for predicting the effects of specimen geometry and loading condition on fatigue strength. *Metals* **2018**, *8*, 811. [CrossRef]
17. Leitner, M.; Vormwald, M.; Remes, H. Statistical size effect on multiaxial fatigue strength of notched steel components. *Int. J. Fatigue* **2017**, *104*, 322–333. [CrossRef]
18. Lanning, D.B.; Nicholas, T.; Palazotto, A. HCF notch predictions based on weakest-link failure models. *Int. J. Fatigue* **2003**, *25*, 835–841. [CrossRef]
19. Bažant, Z.P.; Novák, D. Probabilistic nonlocal theory for quasibrittle fracture initiation and size effect. I: Theory. *J. Eng. Mech.* **2000**, *126*, 166–174. [CrossRef]
20. Makkonen, M. Statistical size effect in the fatigue limit of steel. *Int. J. Fatigue* **2001**, *23*, 395–402. [CrossRef]
21. Härkegård, G.; Halleraker, G. Assessment of methods for prediction of notch and size effects at the fatigue limit based on test data by Böhm and Magin. *Int. J. Fatigue* **2010**, *32*, 1701–1709. [CrossRef]
22. Brown, H.; Mischke, C.; Shigley, J. *Standard Handbook of Machine Design*; McGraw-Hill Education: Raleigh, NC, USA, 2004.
23. Lee, Y.L.; Pan, J.; Hathaway, R.; Barkey, M. *Fatigue Testing and Analysis: Theory and Practice*; Butterworth-Heinemann: Oxford, UK, 2005; Volume 13.
24. Nakai, Y.; Hashimoto, A.; Imanishi, T.; Hiwa, C. Size effect on fatigue strength of metallic micro-materials. In *Proceedings of Asian-Pacific Conference on Fracture and Strength*; Proceedings of Asian-Pacific Conference on Fracture and Strength: Tokyo, Japan, 1999; Volume 99.

25. Biancolini, M.; Brutti, C.; Paparo, G.; Zanini, A. Fatigue cracks nucleation on steel, acoustic emission and fractal analysis. *Int. J. Fatigue* **2006**, *28*, 1820–1825. [CrossRef]

26. Meriaux, J.; Boinet, M.; Fouvry, S.; Lenain, J. Identification of fretting fatigue crack propagation mechanisms using acoustic emission. *Tribol. Int.* **2010**, *43*, 2166–2174. [CrossRef]

27. Lin, L.; Chu, F. HHT-based AE characteristics of natural fatigue cracks in rotating shafts. *Mech. Syst. Signal Process.* **2012**, *26*, 181–189. [CrossRef]

28. Meriaux, J.; Fouvry, S.; Kubiak, K.; Deyber, S. Characterization of crack nucleation in TA6V under fretting–fatigue loading using the potential drop technique. *Int. J. Fatigue* **2010**, *32*, 1658–1668. [CrossRef]

29. Todoroki, A.; Mizutani, Y.; Suzuki, Y.; Haruyama, D. Fatigue damage detection of CFRP using the electrical resistance change method. *Int. J. Aeronaut. Space Sci.* **2013**, *14*, 350–355. [CrossRef]

30. Zilberstein, V.; Schlicker, D.; Walrath, K.; Weiss, V.; Goldfine, N. MWM eddy current sensors for monitoring of crack initiation and growth during fatigue tests and in service. *Int. J. Fatigue* **2001**, *23*, 477–485. [CrossRef]

31. Grammatikos, S.; Kordatos, E.; Matikas, T.; Paipetis, A. Real-time debonding monitoring of composite repaired materials via electrical, acoustic, and thermographic methods. *J. Mater. Eng. Perform.* **2014**, *23*, 169–180. [CrossRef]

32. ToolBox, E. Beams-Fixed at Both Ends-Continuous and Point Loads. 2004. Available online: https://www.engineeringtoolbox.com/beams-fixed-both-ends-support-loads-deflection-d_809.html (accessed on 20 June 2019).

33. ToolBox, E. Beams-Supported at Both Ends-Continuous and Point Loads. 2009. Available online: https://www.engineeringtoolbox.com/beam-stress-deflection-d_1312.html (accessed on 20 June 2019).

34. ASTM. *B593-96(2014)e1 Standard Test Method for Bending Fatigue Testing for Copper-Alloy Spring Materials*; ASTM International: West Conshohocken, PA, USA, 2009. [CrossRef]

35. De, P.; Obermark, C.; Mishra, R. Development of a reversible bending fatigue test bed to evaluate bulk properties using sub-size specimens. *J. Test. Eval.* **2008**, *36*, 402–405.

36. Gohil, P.; Panchal, H.N.; Sohail, S.M.; Mahant, D.V. Experimental and FEA Prediction of Fatigue Life in Sheet Metal (IS 2062). *Int. J. Appl. Res. Stud.* **2013**, *MH 1*, 1.

37. Haidyrah, A.S.; Castano, C.; Newkirk, J.W. An experimental study on bending fatigue test with a krouse-type fatigue specimen. In Proceedings of the 2014 ANS Winter Meeting and Nuclear Technology Expo, Anaheim, CA, USA, 9–13 November 2014; pp. 1–4.

38. Haidyrah, A.S.; Newkirk, J.W.; Castaño, C.H. Characterization a bending fatigue mini-specimen technique (krouse type) of nuclear materials. In *TMS 2015 144th Annual Meeting & Exhibition*; Springer: Berlin, Germany, 2015; pp. 1225–1232.

39. Haidyrah, A.S.; Newkirk, J.W.; Castaño, C.H. Weibull statistical analysis of Krouse type bending fatigue of nuclear materials. *J. Nucl. Mater.* **2016**, *470*, 244–250. [CrossRef]

40. Beer, F.P.; Johnston, E.R.; DeWolf, J.T.; Mazurek, D.F. *Statics and Mechanics of Materials*; McGraw-Hill Education: New York, NY, USA, 2017.

41. Thomson, W. *Theory of Vibration with Applications*; CRC Press: Boca Raton, FL, USA, 2018.

42. Wang, X.; Chung, D. Real-time monitoring of fatigue damage and dynamic strain in carbon fiber polymer-matrix composite by electrical resistance measurement. *Smart Mater. Struct.* **1997**, *6*, 504. [CrossRef]

43. Stephens, R.I.; Fatemi, A.; Stephens, R.R.; Fuchs, H.O. *Metal Fatigue in Engineering*; John Wiley & Sons: Hoboken, NJ, USA, 2000.

44. Mughrabi, H. Microstructural mechanisms of cyclic deformation, fatigue crack initiation and early crack growth. *Philos. Trans. R. Soc. A Math. Phys. Eng. Sci.* **2015**, *373*, 20140132. [CrossRef] [PubMed]

45. Olakanmi, E.O.; Cochrane, R.; Dalgarno, K. A review on selective laser sintering/melting (SLS/SLM) of aluminium alloy powders: Processing, microstructure, and properties. *Prog. Mater. Sci.* **2015**, *74*, 401–477. [CrossRef]

46. Zhang, B.; Li, Y.; Bai, Q. Defect formation mechanisms in selective laser melting: A review. *Chin. J. Mech. Eng.* **2017**, *30*, 515–527. [CrossRef]

47. Karnati, S.; Axelsen, I.; Liou, F.; Newkirk, J.W. Investigation of tensile properties of bulk and SLM fabricated 304L stainless steel using various gage length specimens. In Proceedings of the 27th Annual International Solid Freeform Fabrication Symposium—An Additive Manufacturing Conference, Austin, TX, USA, 8–10 August 2016; pp. 592–604.

48. Al-Shahrani, S.; Marrow, T. Effect of Surface Finish on Fatigue of Stainless Steels. In Proceedings of the ICF12, Ottawa, ON, Canada, 14 October 2009.

Article

Effects of Ultrasonic Impact Treatment on the Stress-Controlled Fatigue Performance of Additively Manufactured DMLS Ti-6Al-4V Alloy

Peter Walker [1], Sinah Malz [2], Eric Trudel [1], Shaza Nosir [1], Mostafa S.A. ElSayed [1,*] and Leo Kok [3]

[1] Department of Mechanical and Aerospace Engineering, Carleton University, Ottawa, ON K1S 5B6, Canada; PeterMWalker@cmail.carleton.ca (P.W); EricTrudel@cmail.carleton.ca (E.T); ShazaNosir@cmail.carleton.ca (S.N.)

[2] Department of Mechanical Engineering, University of Rostock, 18051 Rostock, Germany; sinah.malz@uni-rostock.de

[3] Department of Advanced Structures, Bombardier Aerospace, Toronto, ON M3K 1Y5, Canada; leo.kok@aero.bombardier.com

* Correspondence: mostafa.elsayed@carleton.ca; Tel.: +1-613-520-2600 (ext. 4138)

Received: 10 September 2019; Accepted: 18 October 2019; Published: 8 November 2019

Abstract: Additive manufacturing (AM) offers many advantages for the mechanical design of metal components. However, the benefits of AM are offset to a certain extent by the poor surface finish and high residual stresses resulting from the printing process, which consequently compromise the mechanical properties of the parts, particularly their fatigue performance. Ultrasonic impact treatment (UIT) is a surface modification process which is often used to increase the fatigue life of welds in ship hulls and steel bridges. This paper studies the effect of UIT on the fatigue life of Ti-6Al-4V manufactured by Direct Metal Laser Sintering (DMLS). The surface properties before and after the UIT are characterized by surface porosity, roughness, hardness and residual stresses. Results show that UIT enhances the fatigue life of DMLS Ti-6Al-4V parts by suppressing the surface defects originating from the DMLS process and inducing compressive residual stresses at the surface. At the adopted UIT application parameters, the treatment improved the fatigue performance by 200%, significantly decreased surface porosity, reduced the surface roughness by 69%, and imposed a compressive hydrostatic stress of 1644 MPa at the surface.

Keywords: fatigue life improvement; materials characterization; additive manufacturing; ultrasonic impact treatment; DMLS

---

1. Introduction

Additive Manufacturing (AM) offers great promise to the medical [1], aerospace [2], automotive and defense fields [3–5]. It provides the advantage of building complex geometries by fabricating 3D objects one layer at a time using rendered CAD models [6], as a result, several techniques of AM have been developed, including, Electron Beam Melting (EBM) [7], Selective Laser Sintering (SLS) [8], Selective Laser Melting (SLM) [9], and Direct Metal Laser Sintering (DMLS) [10]. Another advantage of AM lies within the wide range of materials that can be manufactured such as plastics [11], metals [3,12–14], ceramics [15], concrete [16] and fiber reinforced polymers [17], among others. The AM of metals are of particular interest to the production of dental implants, aerospace components, and automotive structures [18]. Today, stainless steel [19,20], nickel alloys [21,22], aluminum alloys [23,24] and titanium alloys [25,26] are common materials for metals AM.

DMLS is one of the most common AM processes for 3D printed metals because it maintains dimensional control while producing complex features at high resolution [3]. There are two primary

---

methods of manufacturing by DMLS: powder bed and powder deposition [27]. Powder bed methods rely on a high energy source, typically a laser (although some similar systems use electron beams), to locally sinter or melt metal particles on a powder platform. New powder layers are periodically added while the platform is levelled to accommodate the addition of new material. The 3D model is constructed in a single vertical direction [6]. Alternatively, powder deposition directly deposits the metal powder and melts it in place using a high-powered laser [28]. Unlike the powder bed method, which is typically restricted to one type of alloy, powder deposition has the ability to include different metal powders for functionally graded materials [29].

This process however, also has a number of drawbacks. Parts produced via DMLS typically have poorer mechanical properties compared to those produced by traditional means, which has relegated the potential uses to prototypes and short-term tooling operations [30]. Incomplete powder melting often leads to very rough surface finish and porosity, which in addition to being aesthetically displeasing also compromises fatigue life and can be a significant issue for wetted surfaces of air and water craft [29,31]. In addition, the rapid heating and effective quenching of the metal results in a highly martensitic microstructure in most alloys [32]. While this produces a material with a high yield strength, it is also intensely brittle [32]. Though the microstructure resists the formation of cracks, once the cracks themselves form, the propagation of the cracks is quite rapid [33]. The high temperature gradient involved in these processes frequently causes thermal stress that compromise the fatigue performance of metals produced by DMLS techniques.

Titanium alloys such as Ti-6Al-4V are commercially available for additive manufacturing. Due to its high strength to weight ratio and fracture toughness, it is an ideal alloy for a wide range of applications in the aerospace and biomedical engineering fields [34]. For instance, Ti-6Al-4V alloy is used in dental laboratories for medical implants and prosthetics due to its corrosion resistance, high specific strength as well as its biocompatibility characteristics. AM also allows for the creation of porous titanium structures that help facilitating bone ingrowth and adhesion for implants [35]. Advances in topology optimization allows for hyper-efficient geometries to be produced exclusively through additive manufacturing. One notable example includes the Airbus A320 nacelle hinge bracket which was eventually produced by AM [36]. With the many advancements in the field of AM, the production of titanium parts has become economically viable. Hence it is imperative that the fatigue properties are improved for the next generation of additively manufactured titanium components [33,37–40]. In the case of Ti-6Al-4V, the hexagonal close-packed (HCP) α phase and trace amounts of a body centered cubic (BCC) β phase are almost entirely replaced with the martensitic α' phase. The poor surface finish and high porosity are also factors that offset the microstructural characteristics on the fatigue performance.

The demand for functional AM parts has been rising as high reduction in assembly costs are available coupled with decreases in mass. One example is the fuel nozzles for the General Electric (GE) LEAP engine have been additively manufactured to be 25% lighter while eliminating previous models that required laborious assembly. Its successful design has now been 3D printed more than 30,000 times since its conception [41]. With the increasing demand for metal AM, researchers have begun to develop techniques to improve the fatigue performance by improving the surface finish and by inducing compressive residual stresses at the surface. Shot peening [42], Ultrasonic Nanocrystal Surface Modification (UNSM) [43–46] and grit blasting [47] are well known examples of beneficial treatments on AM of metals. However, none of these processes were able to address all the following simultaneously: surface roughness, surface porosity, fatigue life and tensile residual stresses.

Studies have shown that Ultrasonic Impact Treatment (UIT) improves the surface finish and fatigue properties in the field of post-welding [48]. UIT is a process in which an indenter vibrating at ultrasonic frequencies slides over a surface. This treatment plastically deforms the surface, improving the surface finish while inducing and redistributing residual stress in the part resulting in enhanced fatigue life [48]. UIT devices operate by inducing plastic deformation from the indenter or impact needle by first exciting a transducer by a controlled voltage input. The power source directly controls the oscillations exhibited by the transducer, which sends its frequencies to the sonotrode (ultrasonic horn) [48]. The research of E. Statnikov et al. [49] compared a variety of methods that improve the fatigue life of welded joints. Other similar methods include Hammer Peening, Shot Peening and Tungsten Inert Gas (TIG) dressing. An improvement of 65% was observed in the UIT joints. In B.N. Mordyuk et al. [50] investigated the enhancements that occur in the surface layer of ultrasonic impacted specimens. It was concluded that the compressive residual stresses and work hardening of the surface layer attributed to the improvement in fatigue properties of processed specimens. A.I. Dekhtyar et al. [51] concluded that at high stress levels, UIT-processed Ti-6Al-4V has a fatigue life that is twice of the pristine (untreated) samples, and a roughness reduction Ra of 75%. While UIT is widely used in fatigue improvement of welded joints, the aim of this paper is to treat the surface of Ti-6Al-4V specimens produced by DMLS using UIT to improve roughness, increase hardness and induce surface compressive residual stresses to enhance the fatigue life of the components.

It is well known that the resistance to tensile fatigue of a metal will increase by the addition of compressive stress [37]. UIT compares favorably to other surface treatments due to its higher impact forces [48]. Quantifying this level of stress by the cold working of the surface for titanium alloys can prove to be beneficial in future engineering applications, as evaluation of impact forces between the pins and the metal surface of UIT is under researched. Force estimations for other cold hardening treatments such as shot peening have been measured. One method used acoustic emissions for velocities of 30 to 88 m/s; however, this was not able to properly determine the impact force [52]. Many other techniques have been developed for measuring shot peening impact forces, which include a shot peening intensity detector; however, most methods cannot be directly applied to measure forces for an ultrasonic impact device [53]. Research into the optimization of UIT parameters has led to the use of oscilloscopes for defining the impact characteristics between the pin and the metal surface [48,54]. The frequencies were able to illustrate the ultrasonic deformations and elastic recovery of the process. The frequency patterns are also able to show the stochastic nature of the impacts that are influenced by both the impact depth and plastic deformation. An additional experiment presented in this paper is conducted to quantify impact forces during UIT on DMLS printed titanium specimens.

2. Materials and Methods

2.1. Specimens Manufacturing

Flat dog-bone fatigue specimens were manufactured in accordance to ASTM E466-15 [55] for force controlled fatigue testing. Figure 1a illustrates the dimensions of the dog-bone in millimeters. The titanium dog-bone specimens were built using Ti-6Al-4V grades 23 powder from AP&C, composed of between 5.5 and 6.75 wt% aluminum, 3.5 and 4.5 wt% vanadium and <0.25 wt% iron and trace amounts of <1 wt% impurities such as oxygen, nitrogen and hydrogen with the balance being titanium [56]. The 3D printer used to manufacture the specimens was an EOSINT M290/400W machine following a general process parameter of layer thickness of 30 μm and volume rate of 5 mm^3/s where the volume rate is a measure of the build speed during laser exposure of the skin area [57]. The process parameters are optimized in such a way as to provide mechanical properties comparable to other literature [58]. The particle size ranged from 15 to 23 μm and the printing layers thickness were 60 μm. The specimens were heat treated in accordance to AMS 2801 to relieve the residual stress induced by the rapid melting and solidification that takes place during the printing process [26,59]. Figure 1b shows the printing orientation, where the printing platform lies in the x–y plane. A single 2D layer is formed on the platform when the laser beam sinters particles starting from the left end and moving towards the right end of the dog-bone shown in Figure 1.

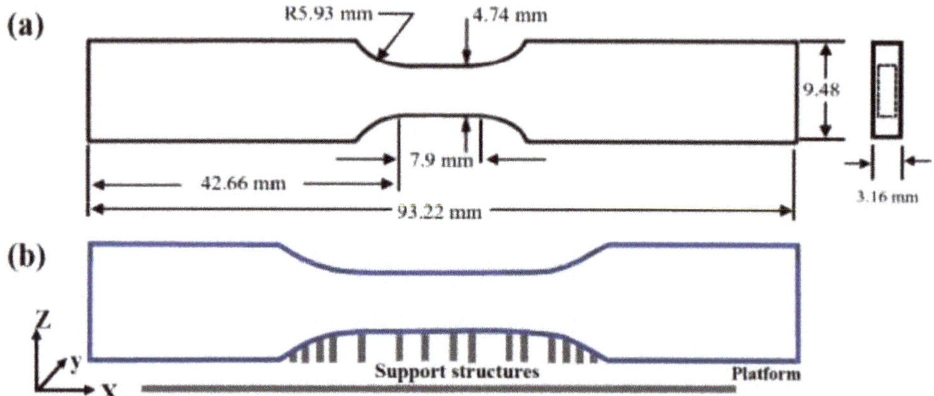

Figure 1. Specimens design and preparation. (**a**) Schematic of flat sheet fatigue specimen with rectangular cross-section; (**b**) orientation of the specimens during DMLS manufacturing process.

When the layer is completed, the platform is levelled in the z-axis allowing a new layer to be added on top of the previous one. Wire Electrical Discharge Machining (EDM) is a precise cutting technique that minimizes the need for excessive post processing machining and was used to breakaway support and remove the specimens from the platform [60]. Lastly, the edges were then polished using an emery cloth of grades 120 and 220 [61].

2.2. UIT Device

The UIT device, displayed in Figure 2, is a 20k Ultrasonic Impact Treatment device, DW-CJ20-1000 produced by Dowell Ultrasonics [62], typically used as a hand-held tool for post-welding processing. It consists of a power supply, shown in Figure 2a, the UIT tool, shown in Figure 2b, the impactor head, shown in Figure 2c and the ultrasonic generator, shown in Figure 2d. It is equipped with slots for four impactors, but for the purposes of this experiment—only one impactor was used, as illustrated in Figure 2c. To provide automated control, a custom-built fixture is used to attach the device to the spindle of a Computer Numerical Control (CNC) Mill. Figure 2e shows the treatment path programmed to minimize the surface roughness while obtaining a uniformly deformed surface. The scanning speed of the CNC machine was set to 1000 mm/min. The spacing between the scans is known as an interval. Amplitude control of the device is controlled by a Fagor 8035M controller [63]. A constant amplitude of 57% of 40 μm was used during treatment as testing showed that this amplitude provided the most consistent plastic deformation of the surface without damaging the samples. The interval for treating titanium alloys typically ranges between 10 and 70 μm [43–46,64]. The path contours were chosen to increase outwards at 71.1 μm intervals to match previous efforts and research into surface treatments on titanium [43,44,65].

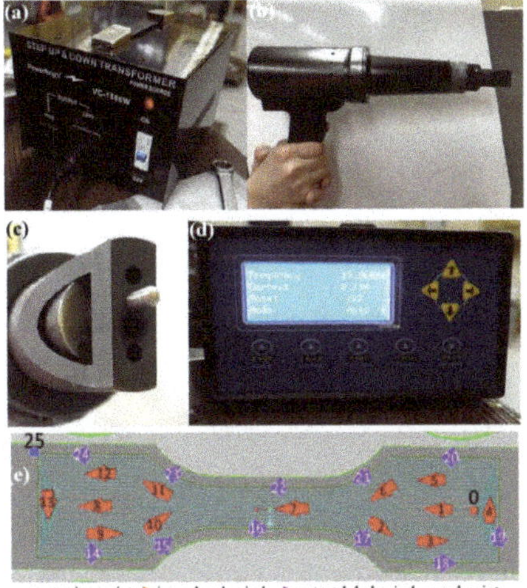

Figure 2. UIT system and application pattern. (**a**) Power supply unit; (**b**) side view of the UIT device; (**c**) front view of the UIT device showing impactor position; (**d**) ultrasonic generator; (**e**) schematic of treatment path.

2.3. Specimens Fixture

To apply the UIT a new fixture was designed, as shown in Figure 3. The Ti-6Al-4V specimens were clamped to an aluminum plate that was supported by four steel rods and four aluminum sleeves, as illustrated in Figure 3. The plate can freely slide along the rods and its motion range is limited to four high precision springs from McMaster Carr [66] placed between the plate and the end of the supporting rods. Compressing the springs allows for a constant static force to be applied onto the samples during treatment. By pushing the UIT device into the plate and compressing the springs a certain distance, the amount of static force can be determined. Aluminum cutting fluid [67] was used to lubricate the rods so that the plate could freely move by the springs. For this experiment, a static force of 30 N was applied at the tool specific constant frequency of 19.86 kHz.

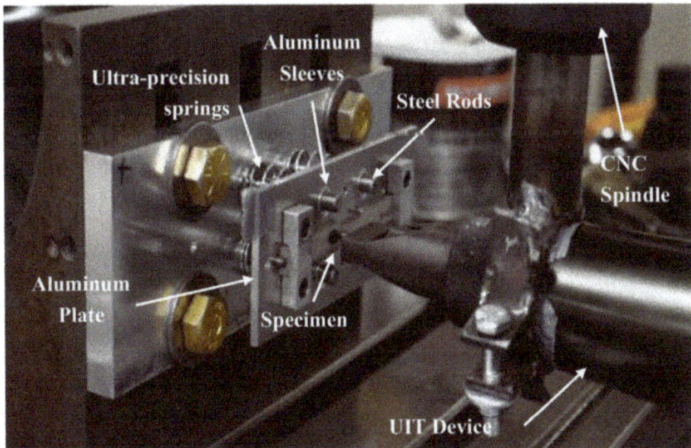

Figure 3. Spring assembly fixture with a clamp mounted to the CNC machine.

2.4. Fatigue Testing

Force controlled fatigue tests were undertaken with a servo-hydraulic 810 Material Testing System (maximum load 100 kN) [68]. The MTS consists of an upper and a lower clamping grip. For consistent and precise alignment, a fence was attached to each grip. All specimens were tested under a clamping pressure of 5.52 MPa, frequency of 25 Hz, maximum stress level of 400 MPa and mean stress level of 200 MPa.

2.5. Microscopy

Microscopic observations were conducted using a combination of optical and scanning electronic microscopy. The former was performed with an Inverted Trinocular Metallurgical Microscope including an AmScope 18 MP MU 1803 Camera [69]. Electron microscopy was performed with a Zeiss GeminiSEM 500 at the University of Ottawa's Centre for Photonics Research [70].

2.6. Roughness

Roughness measurements were performed using DektakXT Stylus Profiler by Bruker [71]. A Peak and Valley analysis was conducted to determine the relative roughness of each sample. As illustrated in Figure 4, a lateral and a longitudinal line scan is performed on each side of a specimen. The scans intersect to form a cross. Roughness measurements were taken over 2 mm in each direction and the results are averaged.

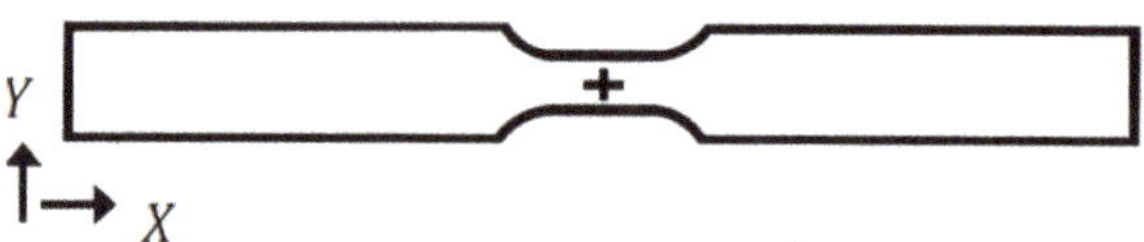

Figure 4. Orientation of line scans for surface roughness.

2.7. Hardness

Rockwell hardness was used to analyze the difference in hardness between treated and untreated specimens following the testing standards of ASTM E18-18a [72]. A Rockwell C test was performed using a test force of 150 kgf on treated and untreated specimens using "The Portable Rockwell Hardness Tester" by Bowers [73].

2.8. Residual Stresses by X-ray Diffraction

Residual stress is defined as the stress remaining in a solid material after an applied force or plastic deformation has taken place. UIT imposes a high plastic strain at the surface of the treated surface, which results in compressive residual stresses. X-ray diffraction (XRD) is a non-destructive method for analyzing the residual stress in a material and is the method of choice for this paper.

Chemical etching of the surface was applied to remove suspected amorphous or oxide material layers, potentially caused by initial stress relaxation treatment. The effect of etching the surface and XRD quality of the scans is illustrated in Figure 5. Peak shapes become much clearer and better defined. It is also important to note that mechanical polishing or grinding will cause lattice strains to be formed at the surface and are not recommended for cleaning samples for residual stress measurements [74]. To conserve and reveal the surface stress layer, a chemical etchant—Kroll's Reagent—was used to remove small amounts of material from the top of the treated and untreated titanium specimens [75,76]. The amount of material removed from the surface was measured to be on average ~ 8 μm.

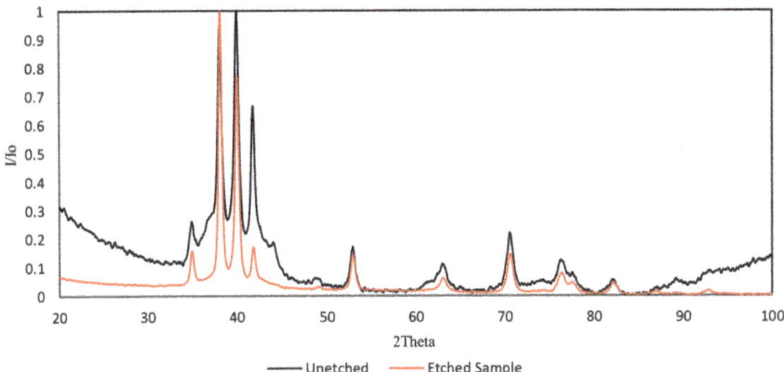

Figure 5. XRD Comparison between etching and without etching.

It is suggested by P. Mercelis et al. [40] that the surface porosity of DMLS manufactured parts poses difficulties for measuring residual stresses due to the presence of zero-stress porosity borders. Stress discontinuities on the surface result in lower residual stresses to be measured than in reality. Hence, cleaning the surface of the samples through etching would also help to reduce the effects of roughness and porosity that may negatively affect the XRD measurements.

The XRD measurements were taken using a Malvern Panalytical Emperyean [77], shown in Figure 6b. The machines power was set to 40 kV with a current of 40 mA for a strong signal response—particularly at higher rocking angles. A half degree diffraction slit was used with a mask size of 2 mm and an anti-scatter slit of 2°. The diffraction arm was a Branson Bragg attachment with a 0.04 mm slit. The incident arm was equipped with a 0.04 mm slit and a 9.1 mm opening.

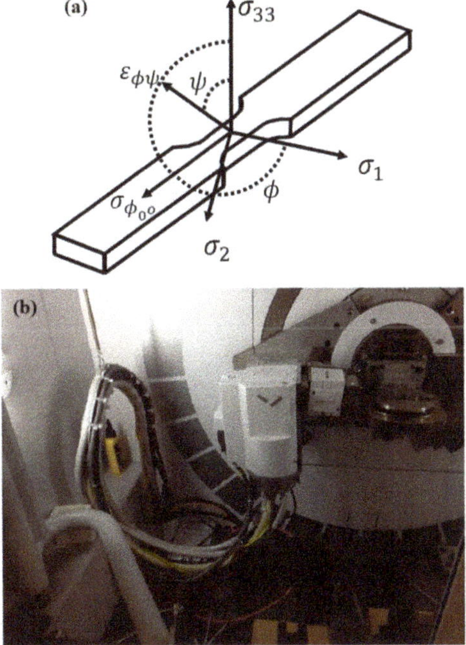

Figure 6. XRD System. (a) Drawing of diffraction directions along the surface and at angles ϕ and ψ. Both σ_1 and σ_2 are perpendicular and reside in the plane of the specimen surface; (b) sample oriented at 0° on multi-purpose stage of Emperyean system.

The analysis method for computing the stress tensor was the Winholtz-Cohen Least squares Analysis [78]. A total of 36 measurements per sample were used for the analysis, each representing a unique tilt and azimuth angles. The chosen x-ray elastic constants of the DMLS Ti-6Al-4V specimens where 2.355×10^{-5} MPa^{-1} as the S_2 constant and -2.9877×10^{-6} MPa^{-1} for the S_1 constant [65,79]. These constants are the interplanar properties for the bulk alpha phase of the metal alloy. The X-ray wavelength was set to 1.519 Å.

The specimen coordinate system had the azimuth or phi (ϕ) of value zero lined up with the horizontal direction (X) as illustrated by the cross seen in Figure 6a. The tilt angles or psi (ψ) is shown in Figure 6a where the 0° begins at 90° from the samples surface.

2.9. Estimation of Impact Force

A piezoelectric force sensor was used to estimate the impact forces at the surface during treatments. To measure the reaction forces at the surface; an alternative fixture was developed to house the sensor and impact plates. This fixture contains linear ball bearings instead of steel bushings to guide the plate, as seen in Figure 7a. The support rods were thickened, and the four previous compression springs were replaced with two larger springs. Proper calibration using a strain gauge was performed so that the correct amount of static force could be determined based on the compression of the spring system. The sensor was attached to the moving plate on the fixture with a titanium specimen used as an impact cap. Operation of the UIT during the estimation of impact force was controlled in much of the same way as during the UIT. The sensor would register the forces from the UIT impacts on the titanium cap. For simplistic design, the cap was a repurposed fatigue specimen.

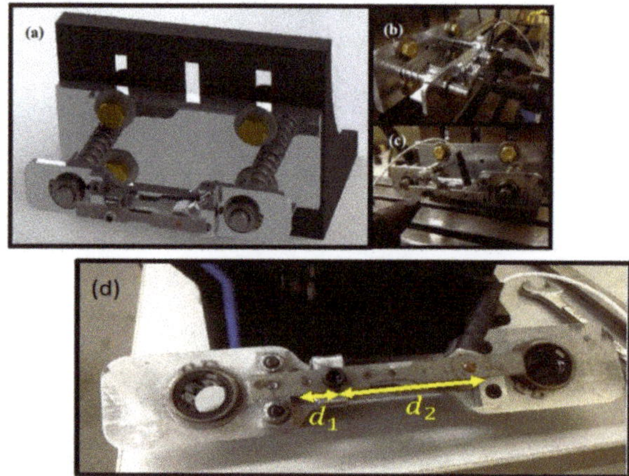

Figure 7. Experimental setup for impact force estimation. (**a–c**) Impact force fixture; (**d**) impact areas and distances.

Due to the design of a repurposed fatigue specimen as an impact surface, reaction forces can be registered by the sensor depending on its distance from the impact. The titanium cap was treated as a double supported beam with one end constrained by a bolt and the other end constrained by the sensor using a double-sided thread. The double supported beam assumption allows for a simple ratio to be developed based equilibrium of moments to solve for the actual impact forces. A diagram of the impact set up can be seen in Figure 7b,c The selected distances between the sensor and the impact region are displayed in Figure 7d, where d_1 is 8.89 mm and d_2 is 47.752 mm. The relationship between the sensor reaction force and the impact force is presented in Equation (1).

$$F_{impact} = -F_{Sensor} \cdot \frac{d_2}{d_1},$$

(1)

where F_{impact} is the force of impact and F_{sensor} is the force registered by the sensor.

Data acquisition was performed using a Lecroy WaveSurfer 3000 Oscilloscope [80]. The force sensor was a PCB piezoelectric force sensor model 208C05. The force range of the sensor was 0 to 4500 kgf. The conversion from voltage to force was assumed to be linear with a sensitivity of 0.2170 mV/N (±15%).

Calibration of the sensor was checked using test procedure AT501-5. The amplifier gain of the sensor was set to 100. Additional shrink tubing was added between the sensor and the wire nuts to prevent vibrations during testing from unscrewing and ejecting the wires from the sensor. The peak voltage read by the sensor for a short ultrasonic impulse was regarded as the impact force for a given static load. The compression of the springs drives the static load with the total stiffness of the system equal to 3.06 N/mm.

2.10. Microstructural Analysis

The effect of the UIT on the microstructure of the material was investigated through Light Optical Microscopy (LOM) and Scanning Electron Microscopy (SEM), the latter being done with both secondary electron and backscatter electron (BSE) detection. Metallographic samples were cut from tensile specimens and mounted in a resin made from phenocure combined with ~15% Technotherm conductive powder for improved SEM performance [81]. Samples were polished with an Allied High Tech MetPrep 3 system with a PH-3 powerhead [82], using SiC paper with a CAMI Grit designation ranging from 120 to 1200 [82], followed by a polish with a suspension of 30% 0.3 µm alumina particles and a final polish of 30% H_2O_2 and 30% 0.05 µm particles [83]. Samples were etched prior to microscopy, using both classical Kroll's reagent and a 10% HF etchant for better grain boundary definition, in keeping with ASTM Standard E407-99 [84].

3. Results and Discussion

3.1. Fatigue Life

A stress-controlled fatigue test was conducted according to DIN 50100 [85] and based on a logarithmic normal distribution. Table 1 displays the fatigue life of Ti-6Al-4V specimens before and after UIT application. The specimens were tested in a servo-hydraulic 810 Material Test system (MTS) at a peak stress level of 400 MPa and minimum stress of 0 MPa. The average number of cycles for the treated and untreated specimens is compared in Figure 8.

Table 1. Fatigue life of treated and untreated specimens at 400 MPa.

Specimen Number	1	2	3	Average
Untreated	2.39×10^4	2.77×10^4	4.02×10^4	2.90×10^4
Treated	8.47×10^4	8.84×10^4	9.63×10^4	8.97×10^4

Evaluation of the fatigue performance of DMLS Ti-6Al-4V shows that the fatigue life of untreated specimens is 77% lower than handbook values [37]. The results in Table 1 and Figure 8 show that the fatigue life of treated specimens at 400 MPa corresponds to a 200% increase compared with that of untreated specimens. In other words, the lifetime of the treated samples is three times as long as untreated samples. D. Cattoni et al. [47] only achieved a 4% fatigue improvement by blasting Ti-6Al-7Nb, whereas A.I. Dekhtyar et al. [51] prolonged the lifetime of Ti-6Al-4V, manufactured using the cost-effective blended elemental powder metallurgy technique, by two orders of magnitude after applying ultrasonic impact treatment.

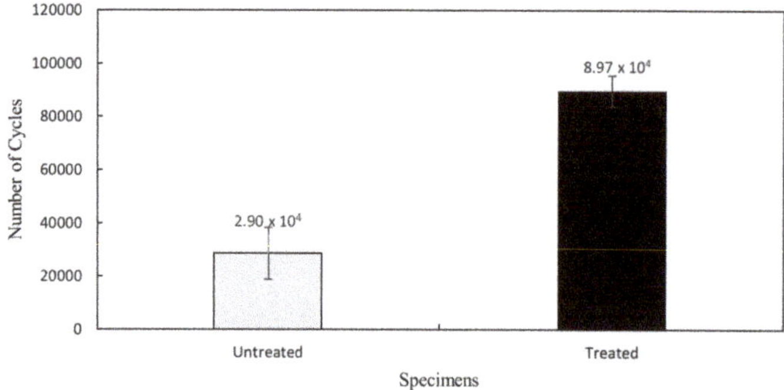

Figure 8. Average fatigue life of treated and untreated Ti-6Al-4V specimens tested at 400 MPa.

3.2. Surface Microscopy

Figure 9a illustrates the surface of Ti-6Al-4V as manufactured by DMLS, which is dominated by the partial melted powder spheres from the additive manufacturing process along with numerous hills and valleys. Despite extensive cleaning in an ultrasonic bath with organic solvents, the rough surface still traps surface contaminants. Tracklines formed by the solidification of the powder bed layers are also visible. These further add to the roughness of the surface and can become sights of crack nucleation and potentially propagation as well [59]. Figure 9b shows the boundary zone between the treated and untreated zones, and the contrast is quite striking. The treated zone is of relatively uniform height, with the spherical powders having been plastically deformed by the impact treatment. However, even this small image of the boundary zone shows that the boundary is quite ragged, with "peninsulas" jutting out from the treated zone and small "islands" of treated area that are separated from the treated zone. The fully treated zone is shown in Figure 9c, and from simple observation it is clear that the surface is much less rough and thus much less conducive to crack nucleation.

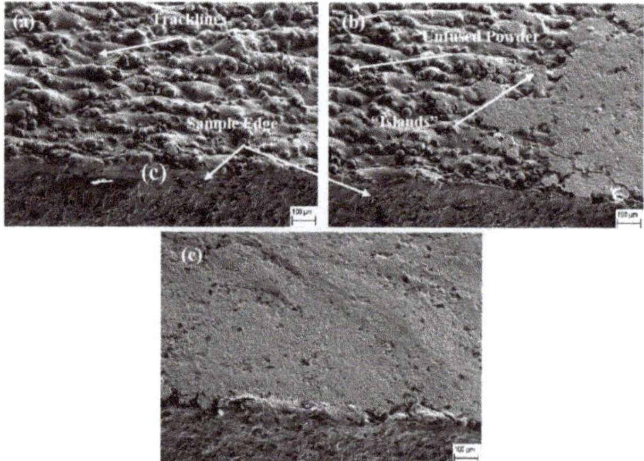

Figure 9. Specimens surface microscopy. (a) Surface of untreated specimen; (b) boundary zone between treated and untreated zone; (c) treated surface of specimen. All images taken with SE2 detector at 15 kV.

The UIT had successfully improved the surface finish of the DMLS Ti-6Al-4V, which corresponds to the improved fatigue performance. Unlike the surface of metals treated by grit blasting and UNSM, the surface of UIT specimens is not jagged. This is due to the combination of vibrating the indenter at ultrasonic speeds while sliding across the surface.

3.3. Roughness

The results from the DektakXT profilometer used to perform a Hills and Valleys profile on both treated and untreated specimens is displayed in Figure 10. The untreated surface is 3.2 times rougher than the treated surface. A control factor for fatigue strength in specimens with high surface roughness is crack propagation, whereas for specimens with low surface roughness it is crack initiation [86]. Additional stress raisers are introduced as a result of rougher surfaces; increasing the number of potential crack initiation sites. Stress raisers as a result of rougher surfaces will reduce crack initiation life and the fatigue limit. Therefore, smooth surfaces can be considered as a contributing factor to the improved fatigue life, as seen by the fatigue performance of the treated samples with improved surface roughness.

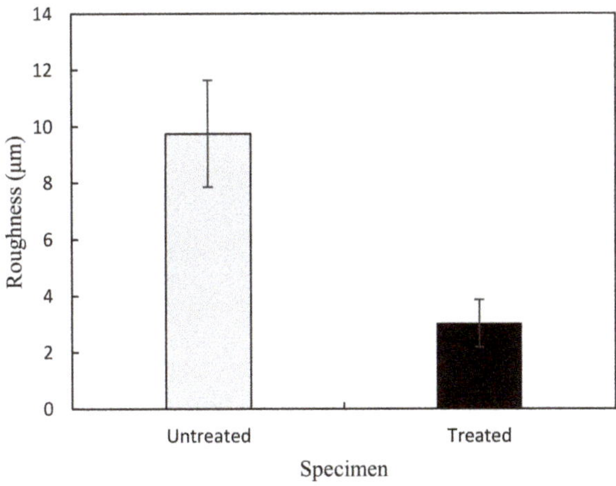

Figure 10. The average roughness of treated and untreated specimens measured by DektakXT.

The effects of UIT on the AM titanium surface are also observed using a two-dimensional map scan as shown in Figure 11. The peaks and valley analysis clearly illustrate the effects on both surface roughness and surface porosity. The untreated surface has a much higher density of hills and valleys unlike the treated surface, which has a much more uniform surface and less pronounced differences between peaks and valleys. The highly irregular surface of the DMLS titanium is attributed to the manufacturing process and incomplete fusion. The final layer of metal powder during printing is not completely melted and produces a very rough surface. The high impact forces of the UIT helps diminish these effects and improves the surface quality and fatigue performance.

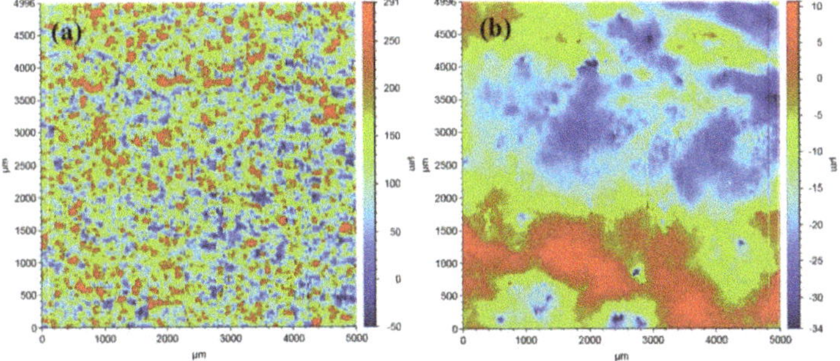

Figure 11. Roughness map scans between the (a) untreated and (b) treated surfaces.

3.4. Hardness

The results from Rockwell C tests performed using a test force of 150 kgf are illustrated in Figure 12, the treated specimens are on average 21% harder than the untreated specimens. However, it is important to note that the hardness is highly variable in both samples. The lowest hardness values from the treated sample are roughly equivalent to the hardest values from the untreated sample. This variability is likely due to the very rough surfaces of both samples. As a result, no statistical conclusion can be drawn from the hardness data about which sample is harder overall. It can be said that the hardness of the treated specimen is more consistent than that of the untreated specimen, and that there are points on the treated specimen that are harder than points on the untreated specimens. This possible increase in hardness at certain points could be attributed to the work hardening done by the UIT, a result of the increase in dislocations density at the surface which will resist further deformation. The increase in hardening could lead to a resistance to crack initiation sites in those locations, which corresponds to an increase in fatigue life [65].

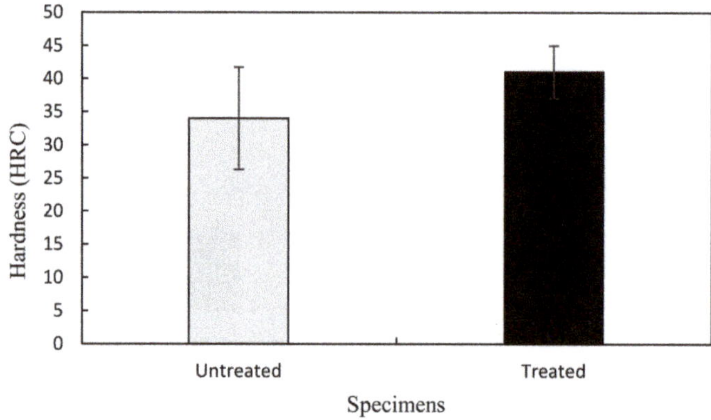

Figure 12. The average hardness of treated and untreated specimens from the Rockwell C test.

3.5. Residual Stresses

For values ranging between 2.5 and 5 μm in depth, residual stresses play a predominant role in improving fatigue life [51]. Based on the results of the residual stress and residual strain measurements, shown in Tables 2 and 3, respectively, the success of the UIT can be attributed to high compressive stresses at the surface of the specimens.

Table 2. Residual stress results (MPa).

Component	Treated—Plane 114		Untreated—Plane 211	
	Stress (MPa)	Error (MPa)	Stress (MPa)	Error (MPa)
σ_{11}	-1.9×10^3	45	-3.0×10^2	20
σ_{22}	-1.8×10^3	45	-2.8×10^2	22
σ_{33}	-1.2×10^3	26	-2.1×10^2	12
σ_{12}	27	6.0	-0.9	2.7
σ_{13}	-1.1	6.0	1.1	2.5
σ_{23}	37	22	19	9.3
σ_{VM}	6.0×10^2	48	84	21
$I_1/3$	-1.6×10^3	68	-2.7×10^2	32

Table 3. Residual strains results.

Component	Treated—Plane 114		Untreated—Plane 211	
	Strain	Error	Strain	Error
ε_{11}	-7.2×10^{-3}	1.8×10^{-4}	-1.1×10^{-3}	7.9×10^{-5}
ε_{22}	-6.7×10^{-3}	1.7×10^{-4}	-9.7×10^{-4}	7.8×10^{-5}
ε_{33}	2.1×10^{-6}	-4.5×10^{-8}	-1.5×10^{-4}	8.8×10^{-6}
ε_{12}	3.2×10^{-4}	7.4×10^{-5}	-1.0×10^{-5}	3.4×10^{-5}
ε_{13}	-1.3×10^{-5}	7.4×10^{-5}	1.2×10^{-5}	3.1×10^{-5}
ε_{23}	4.4×10^{-4}	2.7×10^{-4}	2.2×10^{-4}	1.2×10^{-4}
ε_{VM}	4.7×10^{-3}	5.6×10^{-4}	6.6×10^{-4}	2.4×10^{-4}

These stresses help minimize the generation of tensile stresses at the surface due to cyclic loadings thereby suppressing crack formation and nucleation [49].

The high compressive stresses help minimize the damage caused during cyclic loading by shifting the mean stress downwards. This eventually leads to longer life in the part as compressive stresses will lower the stress ratio. Experimental data has shown that as the stress ratio becomes increasingly negative, longer lives were measured. This can be explained by compressive surface stresses preventing dislocations from moving within the material. The hydrostatic stresses for each tensor reveal the same trend; the untreated sample had a hydrostatic stress of −265 MPa while the treated sample had −1644 MPa. A high compressive hydrostatic stress could also represent an increase in fatigue resistance due to UIT.

3.6. Impact Force Quantification

The impact force with respect to the variation of compressive static force was obtained. The results show that for a static force between 0 and 10 N, the impact force is relatively constant and is on average 20 kN. After which, the force steadily increases up to 72 kN when the static force is equal to 30 N. The trend is slightly parabolic and impact forces at zero newton of static force were measured to be slightly higher than at 10 N of static force. The impact force on the fatigue samples is thus 72 ± 11 kN. For every second, approximately $2.1 \times 10^4 \pm 6.0 \times 10^2$ impacts occur at a static force of near zero, seen in Figure 13. This would mean that for an amplitude of 57% of 40 μm the average absolute vertical speed of the indenter is 0.94×0.03 m/s.

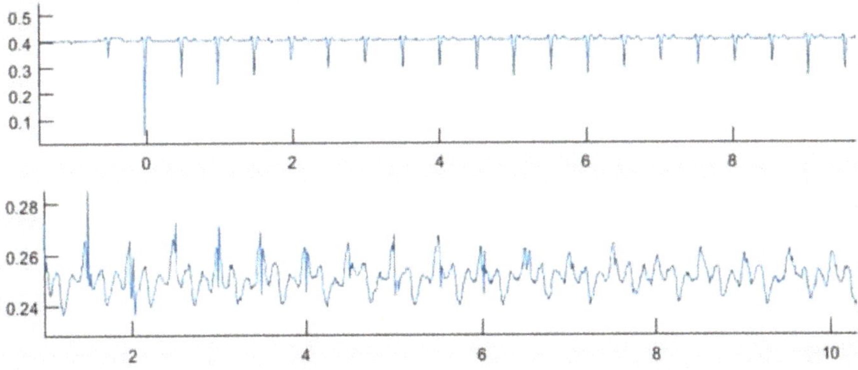

Figure 13. Force vs. time as measured by the force sensor.

The high impact forces are due to metal on metal deformation between the steel pin indenter and the titanium surface. High forces are registered due to the high modulus of elasticity of the mediums and by the ultrasonic vibrations and stress waves [48]. Another factor to consider for high impact forces are due to impulses occurring over a very short time period. The force versus time plots in Figure 13 show the readings of the oscilloscope of a low static force. When minimal static forces are present, the force pattern is rather predictable and consistent. As the static force is increased, the level of stochastic behavior increases dramatically, as seen in Figure 14. The increased randomness is attributed to the increase in plastic deformation which creates widening gaps with varying depth. The subsequent plastic deformation on the surface will then begin to slightly alter the impact forces slightly due to both work-hardening and rebounding forces from the support springs.

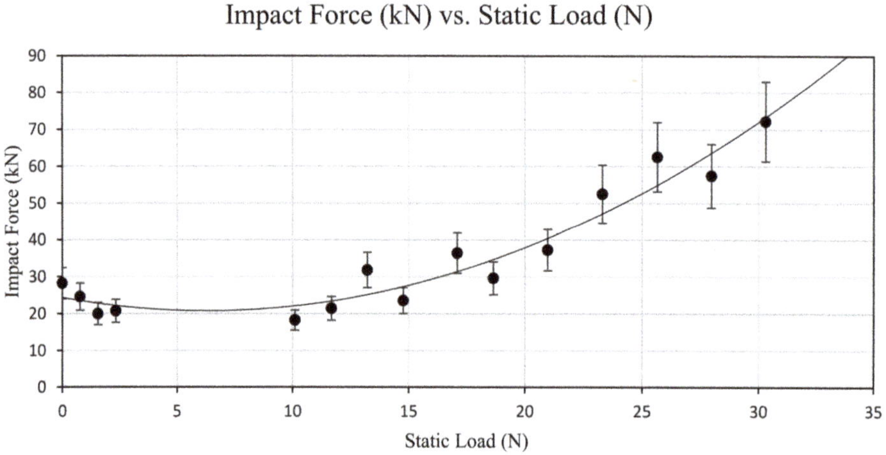

Impact Force (kN) vs. Static Load (N)

Impact Force (kN)

Static Load (N)

Figure 14. Impact force (kN) vs. static load (N).

The impact force on the surface, given one UIT needle, can be replicated if an equivalent force is used to generate the plastic deformation. Hence, other work hardening processes can achieve the same effects of the UIT if similar impact forces are used combined with a transverse velocity along the surface. Possible errors in the impact force assessment can be attributed to off-axis impact forces. Loads applied to the side of the sensor may cause higher than normal readings. This is a result of coupling forces acting on the sensor. However, calibration between the distances of the applied load to sensor was made to ensure that off-axis effects are diminished.

The saturation point of the sensor was reached once the static forces approached 30 N as it was close to the maximum range of the sensor. Any other force impact reading after the saturation point would not have been reliable as the sensor would have been destroyed. It is unknown what impact forces are possible after 30 N as extrapolation of the data is within an area of high uncertainty. The results are also calibrated for the UIT fixtures set up which includes a spring-loaded system. The springs vibrations combined with the indenter's oscillations creates a varied and stochastic process leading to large error bounds during the treatment process. The reaction forces of the springs may also contribute to higher than normal impact forces. Due to the calibration of the sensor in combination with the fixture, there is a level of specificity in the force calibration curve that may limit its applicability for future engineering applications.

3.7. Microstructure Analysis

As mentioned previously, the rapid heating and cooling cycle created by most forms of additive manufacturing processes results in a heavily martensitic microstructure. In the case of Ti-6Al-4V, this is characterized by a so called "basket-weave" pattern. The comparison between the microstructure of conventionally wrought Ti-6Al-4V, shown in Figure 15a, and the microstructure of additively manufactured Ti-6Al-4V, shown in Figure 15b, is quite striking. The basket-weave pattern of the overlapping α′ phase is quite visible, compared to the more randomly orientated microstructure produced by more conventional means.

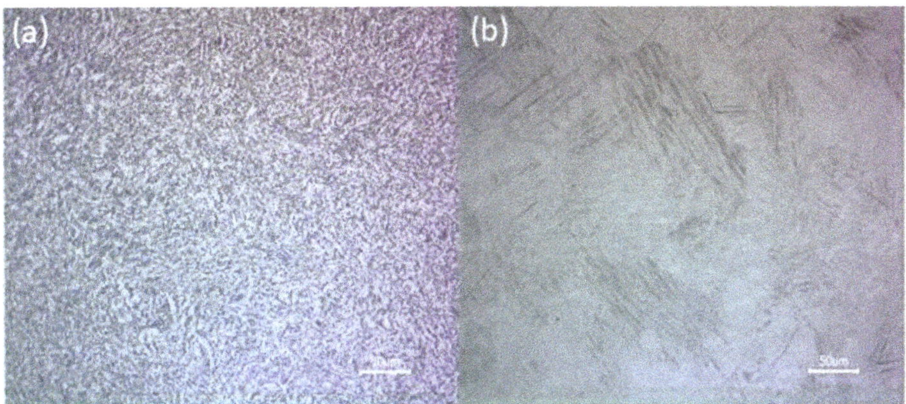

Figure 15. Optical micrographs of (**a**) Wrought Ti-6Al-4V and (**b**) DMLS Ti-6Al-4V.

Neither optical microscopy nor electron microscopy of the treated vs. untreated samples showed a significant change in microstructure. In both cases, the distribution of the α′ and β phases appear randomly distributed, with no obvious relationship between the β phase and the surface of the specimens, as seen in Figure 16a,b. Due to the low temperatures observed during the cold working (the samples were too hot to touch immediately after processing, but did not show any evidence of oxidization) the samples never reached a point where dynamic recrystallization could occur, so microstructural changes would be unlikely.

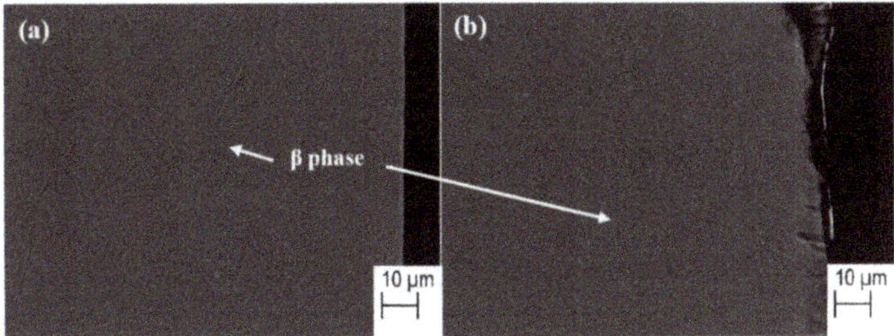

Figure 16. SEM micrographs of (**a**) UIT applied and (**b**) untreated specimens, BSE at 30 kV.

In both cases β phase "needles" were observed at various angles to the surface of the sample, and their lengths were relatively consistent with an average 17.2 μm at the surface in the untreated region and an average 16.8 μm at the surface of the treated region, a 2.3% difference. This suggests that the predominant mechanisms of fatigue life improvement are the addition of compressive residual stress and surface roughness improvements.

4. Conclusions

By applying UIT to the surface of DMLS Ti-6Al-4V, a nearly three-fold increase in fatigue life for cyclic stress levels of 400 MPa was achieved. This improvement can be attributed to the several aspects of the treatment, reduction in surface porosity, a decrease in surface roughness and possibly causing local increases in hardness. By cold working the surface, the barrier for crack nucleation is increased and delaying the onset of fatigue cracking at the surface. The large compressive stresses imposed by UIT on the treated area also suppress crack propagation by offsetting tensile stresses at the surface during cyclic loads. In addition, improved surface finish due to UIT can help to reduce the number of potential sites for crack nucleation by reducing surface porosity, again delaying the onset of crack formation. The impact force of the treatment was determined to be 72 kN and a curve of impact force versus compressive static force was developed. Microstructural observations did not demonstrate any significant change between treated and untreated specimen edges, suggesting the predominant mechanisms of the treatment are the surface modifications and compressive residual stress. In conclusion, UIT can successfully improve the surface finish while simultaneously improve the fatigue life.

Author Contributions: Conceptualization, M.S.A.E., and L.K.; methodology, P.W., S.M., E.T. and S.N.; software, E.T.; validation, E.T., P.W.; formal analysis, E.T. and P.W.; investigation, P.W., S.M., E.T. and S.N.; resources, M.S.A.E.; data curation, E.T, P.W.; writing—original draft preparation, E.T., P.W. and S.N.; writing—review and editing, P.W. and M.S.A.E.; visualization, E.T., P.W. and S.N.; supervision, M.S.A.E.; project administration, M.S.A.E.; funding acquisition, M.S.A.E.

Funding: This research was funded by BOMBARDIER Aerospace of Montreal, in collaboration with CARIC National Forum (grant number CARIC-CRIAQ MDO-1601_TRL4+) and MITACS Canada (grant number IT07461).

Acknowledgments: Mostafa S.A. ElSayed acknowledges the financial support provided by BOMBARDIER Aerospace of Montreal, in collaboration with CARIC National Forum (grant number CARIC-CRIAQ MDO-1601_TRL4+) and MITACS Canada (grant number IT07461). The authors would like to extend their acknowledgements to the employees of the machine shop, and Dix of the X-ray Diffractometry Lab, both at Carleton University for their guidance and patience, the Centre of Photonics Research and Jeff Ovens at the X-ray Core Facility, both at the University of Ottawa for their training and assistance.

References

1. Javaid, M.; Haleem, A. Additive manufacturing applications in medical cases: A literature based review. *Alex. J. Med.* **2017**, *54*, 411–422. [CrossRef]

2. Schiller, G.J. Additive manufacturing for Aerospace. In Proceedings of the 2015 IEEE Aerospace Conference, Big Sky, MT, USA, 7–14 March 2015.

3. Frazier, W.E. Metal Additive Manufacturing: A Review. *J. Mater. Eng. Perform.* **2014**, *23*, 1917–1928. [CrossRef]

4. Kruth, J.-P.; Leu, M.C.; Nakagawa, T. Progress in Additive Manufacturing and Rapid Prototyping. *CIRP Ann.* **1998**, *47*, 525–540. [CrossRef]

5. United States Air Force Global Horizons Final Report: United States Air Force Global Science and Technology Vision. 2013. Available online: https://www.hsdl.org/?abstract&did=741377 (accessed on 8 November 2019).

6. Gebhardt, A.; Jan-Steffen, H.T. *Addititve Manufacturing*; Carl Hanser Verlag GmbH & Co. KG: München, Germany, 2016.

7. Murr, L.E.; Gaytan, S.M.; Ceylan, A.; Martinez, E.; Martinez, J.L.; Hernandez, D.H.; Machado, B.I.; Ramirez, D.A.; Medina, F.; Collins, S.; et al. Characterization of titanium aluminide alloy components fabricated by additive manufacturing using electron beam melting. *Acta Mater.* **2010**, *58*, 1887–1894. [CrossRef]

8. Kruth, J.-P.; Mercelis, P.; Vaerenbergh, V.J.; Froyen, L.; Rombouts, M. Binding mechanisms in selective laser sintering and selective laser melting. *Rapid. Prototype J.* **2005**, *1*, 26–36. [CrossRef]

9. Gorny, B.; Niendorf, T.; Lackmann, J.; Thoene, M.; Troester, T.; Maier, H. In Situ Characterization of the Deformation and Failure Behavior of Non-Stochastic Porous Structures Processed by Slective Laser Melting. *Mater. Sci. Eng. A* **2011**, *27*, 2–7.

10. Ciocca, L.; Fantini, M.; de Crescenzio, F.; Corinaldesi, G.; Scoti, R. Direct metal laser sintering (DMLS) of a customized titanium mesh for prosthetically guided bone regeneration of atrophic maxillary arches. *Med. Biol. Eng. Comput.* **2011**, *49*, 1347–1352. [CrossRef] [PubMed]

11. Ning, F.; Cong, W.; Qiu, J.; Wei, J.; Wang, S. Additive manufacturing of carbon fiber reinforced thermoplastic composites using fused deposition modeling. *Compos. Part B Eng.* **2015**, *80*, 369–378. [CrossRef]

12. Herzog, D.; Seyda, V.; Wycisk, E.; Emmelmann, C. Additive manufacturing of metals. *Acta Mater.* **2016**, *117*, 371–392. [CrossRef]

13. Milewski, J.O. *Additive Manufacturing of Metals: From Fundamental Technology to Rocket Nozzles, Medical Implants, and Custom Jewelry*; Springer International Publishing AG: Cham, Switzerland, 2017.

14. Mueller, B. Additive Manufacturing Technologies—Rapid Prototyping to Direct Digital Manufacturing. *Assem. Autom.* **2012**, *32*. [CrossRef]

15. Deckers, J.; Vleugels, J.; Kruth, J.-P. Additive Manufacturing of Ceramics: A Review. *J. Ceram. Sci. Technol.* **2014**, *5*, 245–260.

16. Bos, F.; Wolfs, R.; Ahmed, Z.; Salet, T. Additive manufacturing of concrete in construction: Potentials and challenges of 3D concrete printing. *Virtual Phys. Prototype* **2016**, *11*, 209–225. [CrossRef]

17. Stansbury, J.W.; Idacavage, M.J. 3D printing with polymers: Challenges among expanding options and opportunities. *Dent. Mater.* **2016**, *32*, 54–64. [CrossRef] [PubMed]

18. Guo, N.; Leu, M.C. Additive manufacturing: Technology, applications and research needs. *Front. Mech. Eng.* **2013**, *8*, 215–243. [CrossRef]

19. Spierings, A.B.; Starr, T.L.; Wegener, K. Fatigue performance of additive manufactured metallic parts. *Rapid Prototype J.* **2013**, *19*, 88–94. [CrossRef]

20. Islam, M.; Purtonen, T.; Piili, H.; Salminen, A.; Nyrhila, O. Temperature Profile and Imaging Analysis of Laser Additive Manufacturing of Stainless Steel. *Phys. Procedia* **2013**, *41*, 835–842. [CrossRef]

21. Kanagarajah, P.; Brenne, F.; Niendorf, T.; Maier, H.J. Inconel 939 processed by selective laser melting: Effect of microstructure and temperature on the mechanical properties under static and cyclic loading. *Mater. Sci. Eng. A* **2013**, *588*, 188–195. [CrossRef]

22. Vilaro, T.; Colin, C.; Bartout, J.D.; Naze, L.; Sennour, M. Microstructural and mechanical approaches of the selective laser melting process applied to a nickel-base superalloy. *Mater. Sci. Eng. A* **2012**, *534*, 446–451. [CrossRef]

23. Schimidtke, K.; Palm, F.; Hawkins, A.; Emmelmann, C. Process and Mechanical Properties: Applicability of a Scandium modified Al-alloy for Laser Additive Manufacturing. *Phys. Procedia* **2011**, *12*, 369–374. [CrossRef]

24. Brandl, E.; Heckenberger, U.; Holzinger, V.; Buchbinder, D. Additive manufactured AlSi10Mg samples using Selective Laser Melting (SLM): Microstructure, high cycle fatigue, and fracture behavior. *Mater. Des.* **2012**, *34*, 159–169. [CrossRef]

25. Hrabe, N.W.; Quinn, T.P.; Kircher, R. Effects of Processing on Microstructure and Mechanical Properties of Ti-6Al-4V Fabricated using Electron Beam Melting (EBM), Part 1: Distance from Build Plate and Part Size. *Mater. Sci. Eng. A* **2013**, *573*, 264–270. [CrossRef]

26. Chan, K.S.; Koike, M.; Mason, R.L.; Okabe, T. Fatigue Life of Titanium Alloys Fabricated by Additive Layer Manufacturing Techniques for Dental Implants. *Metall. Mater. Trans. A* **2013**, *44*, 1010–1022. [CrossRef]

27. Conner, B.P.; Manogharan, G.P.; Martof, A.N.; Rodomsky, L.M.; Rodomsky, C.M.; Jordan, D.C.; Limperos, J.W. Making sense of 3-D printing: Creating a map of additive manufacturing products and services. *Addit. Manuf.* **2014**, *1*, 64–76. [CrossRef]

28. Marcus, H.; Barlow, J.; Beaman, J.; Bourell, D.; Agarwala, M. Direct selective laser sintering of metals. *Rapid Prototype J.* **1995**, *1*, 26–36.

29. Sames, W.J.; List, F.A.; Pannala, S.; Dehoff, R.R.; Babu, S.S. The metallurgy and processing science of metal additive manufacturing. *Int. Mater. Rev.* **2016**, *61*, 315–360. [CrossRef]

30. Metal AM. Applications for metal Additive Manufacturing Technology. Available online: https://www.metal-am.com/introduction-to-metal-additive-manufacturing-and-3d-printing/applications-for-additive-manufacturing-technology/ (accessed on 8 November 2019).

31. Vayre, B.; Vignat, F.; Villeneuve, F. Designing for Additive Manufacturing. *Procedia CIRP* **2012**, *3*, 632–637. [CrossRef]

32. Leuders, S.; Thone, M.; Riemer, A.; Niendorf, T.; Troster, T.; Richard, H.A.; Maier, H.J. On the mechanical behaviour of titanium alloy TiAl6V4 manufactured by selective laser melting: Fatigue resistance and crack growth performance. *Int. J. Fatigue* **2013**, *48*, 300–307. [CrossRef]

33. Santos, E.C.; Abe, F.; Osakada, K.; Kitamura, Y.; Shiomo, M. Mechanical properties of pure titanium models processed by selective laser melting. In Proceedings of the Solid Freeform Fabrication Symposium, Austin, TX, USA, 5–7 August 2002; pp. 180–186.

34. Li, P.; Warner, D.H.; Fatemi, A.; Phan, N. Critical assessment of the fatigue performance of additively manufactured Ti–6Al–4V and perspective for future research. *Int. J. Fatigue* **2016**, *85*, 130–143. [CrossRef]

35. Ashan, M. 3D Printing and Titanium Alloys: A Paper Review. *Eur. Acad. Res.* **2016**, *3*.

36. Tomlin, M.; Meyer, J. Topology Optimization of an Additive Layer Manufactured (ALM) Aerospace Part. In Proceedings of the 7th Altair CAE Technology Conference 2011, Warwickshire, UK, 10 May 2011.

37. Edwards, P.; Ramulu, M. Fatigue performance evaluation of selective laser melted Ti-6Al-4V. *Mater. Sci. Eng. A* **2014**, *598*, 327–337. [CrossRef]

38. Wycisk, E.; Solbach, A.; Siddique, S.; Herzog, D.; Walther, F.; Emmelmann, C. Effects of Defects in Laser Additive Manufactured Ti-6Al-4V on Fatigue Properties. *Phys. Procedia* **2014**, *56*, 371–378. [CrossRef]

39. Li, S.J.; Murr, L.E.; Cheng, X.Y.; Zhang, Z.B.; Hao, Y.L.; Yang, R.; Medina, F.; Wicker, R.B. Compression fatigue behavior of Ti–6Al–4V mesh arrays fabricated by electron beam melting. *Acta Mater.* **2012**, *60*, 793–802. [CrossRef]

40. Mercelis, P.; Kruth, J.-P. Residual stresses in selective laser sintering and selective laser melting. *Rapid Prototype J.* **2006**, *12*, 254–265. [CrossRef]

41. Sher, D. GE Aviation Already 3D Printed 30,000 Fuel Nozzles for its LEAP Engine. 3D Printing Media Network. Available online: https://www.3dprintingmedia.network/ge-aviation-already-3d-printed-30000-fuel-nozzles-for-its-leap-engine/ (accessed on 19 May 2019).

42. Zhan, K.; Jiang, C.H.; Ji, V. Residual Stress Relaxation of Shot Peened Deformation Surface Layer on S30432 Austenite Steel under Applied Loading. *Mater. Trans.* **2012**, *53*, 1578–1581. [CrossRef]

43. Ma, C.; Andani, M.T.; Qin, H.; Moghaddam, N.S.; Ibrahim, H.; Jahadakbar, A.; Amerinatanzi, A.; Ren, Z.; Zhang, H.; Doll, G.L.; et al. Improving surface finish and wear resistance of additive manufactured nickel-titanium by ultrasonic nano-crystal surface modification. *J. Mater. Process. Technol.* **2017**, *249*, 433–440. [CrossRef]

44. Ma, C.; Dong, Y.; Ye, C. Improving Surface Finish of 3D-printed Metals by Ultrasonic Nanocrystal Surface Modification. *Procedia CIRP* **2016**, *45*, 319–322. [CrossRef]

45. Wang, Y.; Sheng, A. *Manufacturing and Engineering Technology*; CRC Press: Boca Raton, FL, USA, 2014; p. 1.

46. Ye, C.; Telang, A.; Gill, A.; Wen, X.; Mannava, S.R.; Qian, D.; Vasudevan, V.K. Effects of Ultrasonic Nanocrystal Surface Modification on the Residual Stress, Microstructure, and Corrosion Resistance of 304 Stainless Steel Welds. *Metall. Mater. Trans. A* **2018**, *49*, 972–978. [CrossRef]

47. Cattoni, D.; Ferrari, C.; Lebedev, L.; Pazos, L.; Svoboda, H. Effect of Blasting on the Fatigue Life of Ti-6Al-7Nb and Stainless Steel AISI 316 LVM. *Procedia Mater. Sci.* **2012**, *1*, 461–468. [CrossRef]

48. Statnikov, E.S.; Korolkov, O.V.; Vityazev, V.N. Physics and Mechanism of Ultrasonic Impact. *Ultrasonics* **2006**, *44*, e533–e538. [CrossRef]

49. Statnikov, E.S.; Muktepavel, V.O.; Blomqvist, A. Comparison of Ultrasonic Impact Treatment (UIT) and Other Fatigue Life Improvement Methods. *Weld. World* **2002**, *46*, 20–32. [CrossRef]

50. Mordyuk, B.N.; Prokopenko, G.I. Ultrasonic impact peening for the surface properties' management. *J. Sound Vib.* **2007**, *308*, 855–866. [CrossRef]

51. Dekhtyar, A.I.; Mordyuk, B.N.; Savvakin, D.G.; Bondarchuk, V.I.; Moiseeva, I.V.; Khripta, N.I. Enhanced fatigue behavior of powder metallurgy Ti–6Al–4V alloy by applying ultrasonic impact treatment. *Mater. Sci. Eng. A* **2015**, *641*, 348–359. [CrossRef]

52. Ikeda, S.; Kobayashi, Y.; Matsui, A. Evaluation of shot peening impact force by AE method. In Proceedings of the 12th International Conference on Shot Peening, Goslar, Germany, 15–18 September 2014; pp. 244–249.

53. DeClark, B.W.; Loersch, J.F.; Neal, J.W.; Weber, J.H. Shot Peening Intensity Detector. U.S. Patent 4,470,292, 11 September 1984.

54. Statnikov, E.S.; Vityazev, V.N.; Korolkov, O.V. Study of comparative characteristics of ultrasonic impact and optimization of deformation treatment processes. In Proceedings of the 5th World Congress on Ultrasonics, Paris, France, 5–7 September 2003.

55. ASTM International. *Standard Practice for Conducting Force Controlled Constant Amplitude Axial Fatigue Tests of Metallic Materials*; ASTM E466-15; ASTM International: West Conshohocken, PA, USA, 2015.

56. OmniSint-160. 2015. Available online: http://lasersint.com/wordpress/?page_id=449 (accessed on 12 September 2019).

57. GE Additive, Ti-6AI-4V Grade 23. 2019. Available online: https://www.ge.com/additive/additive-manufacturing/materials/apc/ti-6al-4v-23 (accessed on 27 August 2019).

58. EOS, EOS M290/400W Standard, AD, WEIL. 2014. Available online: https://www.sculpteo.com/media/imagecontent/EOS-Titanium-Ti64ELI.pdf (accessed on 28 August 2019).

59. Qiu, C.; Adkins, N.J.E.; Attallah, M.M. Microstructure and tensile properties of selectively laser-melted and of HIPed laser-melted Ti–6Al–4V. *Mater. Sci. Eng. A* **2013**, *578*, 230–239. [CrossRef]

60. Dimla, D.E.; Hopkinson, N.; Rothe, H. Investigation of complex rapid EDM electrodes for rapid tooling applications. *Int. J. Adv. Manuf. Technol.* **2004**, *23*, 249–255. [CrossRef]

61. Emery Cloth Sheet 220–240 Grit. 2019. Available online: https://www.stuller.com/products/10-11014/56351/?groupId=190773 (accessed on 27 August 2019).

62. Ultrasonics, D. 20k1200w Ultrasonic Portable Spot Welding Machine. 2019. Available online: http://www.dowellsonic.com/product/20k1200w-ultrasonic-portable-spot-welding-machine.html (accessed on 28 August 2019).

63. CNC Parts Department. Fagor 8035 CNC. Available online: https://www.cncpd.com/product-category/cnc-controllers/fagor-automation/fagor-8035/ (accessed on 28 August 2019).

64. Günther, J.; Krewerth, D.; Lippmann, T.; Leuders, S.; Tröster, T.; Weidner, A.; Biermann, H.; Niendorf, T. Fatigue life of additively manufactured Ti–6Al–4V in the very high cycle fatigue regime. *Int. J. Fatigue* **2017**, *94*, 236–245. [CrossRef]

65. Zhang, H.; Chiang, R.; Qin, H.; Ren, Z.; Hou, X.; Lin, D.; Doll, G.L.; Vasudevan, V.K.; Dong, Y.; Ye, C. The effects of ultrasonic nanocrystal surface modification on the fatigue performance of 3D-printed Ti64. *Int. J. Fatigue* **2017**, *103*, 136–146. [CrossRef]

66. Mc Master-Carr. Compression Springs. 2019. Available online: https://www.mcmaster.com/springs (accessed on 17 August 2019).

67. CIMSTAR®QUAL STAR C Product Information Flyer. 2011. Available online: http://www.cimcool.com/wp-content/uploads/manage-msdspif/pif/CIMSTAR%20QUALSTAR%20C.pdf (accessed on 21 Aug 2019).

68. MTS, MTS 810 & 858 Test Systems. 2006. Available online: https://www.upc.edu/sct/documents_equipament/d_77_id-412.pdf (accessed on 21 July 2019).

69. AmScope, 18MP USB 3.0 Color CMOS C-Mount Microscope Camera with Reduction Lens. 2019. Available online: https://www.amscope.com/cameras/18mp-usb3-0-real-time-live-video-microscope-digital-camera.html (accessed on 19 July 2019).

70. Zeiss, ZEISS GeminiSEM. 2019. Available online: https://www.zeiss.com/microscopy/int/products/scanning-electron-microscopes/geminisem.html (accessed on 18 August 2019).

71. Bruker, The Gold Standard in Stylus Profilometry. 2019. Available online: https://www.bruker.com/products/surface-and-dimensional-analysis/stylus-profilometers/dektak-xt/overview.html (accessed on 15 August 2019).

72. ASTM International. *Standard Test Methods for Rockwell Hardness of Metallic Materials*; ASTM E18-18a; ASTM International: West Conshocken, PA, USA, 2018.

73. Bowers Group. Portable Hardness Tester IPX-360. Available online: https://www.bowersgroup.co.uk/ca/product-range/testing-instruments/portable/ipx-360-portable-hardness-tester-ipx-360.html (accessed on 18 August 2019).

74. Flitzpatrick, M.E.; Fry, A.T.; Holdway, P.; Kandil, F.A.; Shackleton, J.; Suominen, L. Determination of Residual Stresses by X-ray Diffraction. In *Measurement Good Practice Guide No. 52*; National Physical Laboratory: Teddington, UK, 2005; Volume 52, Issue 2, pp. 1–68.

75. Ruud, C.O. A Review of Nondestructive Methods for Residual Stress Measurement. *JOM* **1981**, *33*, 35–40. [CrossRef]

76. Schajer, G.S.; Ruud, C.O. *Practical Residual Stress Measurement Methods*; John Wiley & Sons Ltd.: Hoboken, NJ, USA, 2013.

77. Malvern Panalytical. Empyrean. 2019. Available online: https://www.malvernpanalytical.com/en/products/product-range/empyrean-range/empyrean (accessed on 21 August 2019).

78. Zyl, I.V.; Yadroitsava, I.; Yadroitsev, I. Residual stress in ti6al4v objects produced by direct metal laser sintering. *S. Afr. J. Ind. Eng.* **2016**, *27*, 134–141.

79. Novovic, D.; Dewes, R.C.; Aspinwall, D.K.; Voice, W.; Bowen, P. The effect of machined topography and integrity on fatigue life. *Int. J. Mach. Tools Manuf.* **2004**, *44*, 125–134. [CrossRef]

80. Teledyne LeCroy. WaveSurfer 3000 Oscilloscopes. 2019. Available online: https://teledynelecroy.com/options/productseries.aspx?mseries=466&groupid=234 (accessed on 28 August 2019).

81. ANAMET, Mounting Polymers. 2016. Available online: http://www.anamet.com/media/other/357089-CATALOG_6_mounting_EN.pdf (accessed on 11 July 2019).

82. Allied High Tech Products. Metprep 3™ Grinder/Polisher with Power Head. 2019. Available online: http://www.alliedhightech.com/Equipment/metprep-3-grinder-polisher-with-powerhead (accessed on 11 August 2019).

83. Sigma-Aldrich, Aluminum Oxide. 2019. Available online: https://www.sigmaaldrich.com/catalog/product/sigald/229423?lang=en®ion=CA (accessed on 8 August 2019).

84. American Society for Testing and Materials. *Standard Practice for Microetching Metals and Alloys*; E 407–99; American Society for Testing and Materials: West Conshohocken, PA, USA, 1999.

85. Deutsches Institut Fur Normung. *DIN 50100: Load Controlled Fatigue Testing–Execution and Evaluation of Cyclic Tests at Constant Load Amplitudes on Metallic Specimens and Components*; Deutsches Institut Fur Normung: Berlin, Germany, 2016.

86. Alang, N.A.; Razak, N.A.; Miskam, A.K. Effect of Surface Roughness on Fatigue Life of Notched Carbon Steel. *Int. J. Eng. Technol.* **2011**, *11*, 160–163.

Review

Part Functionality Alterations Induced by Changes of Surface Integrity in Metal Milling Process: A Review

Caixu Yue [1,*], Haining Gao [1], Xianli Liu [1] and Steven Y. Liang [2]

[1] Harbin University of Science and Technology, Harbin 150080, China; hngao@hrbust.edu.cn (H.G.); Xianli.liu@hrbust.edu.cn (X.L.)

[2] Georgia Institute of Technology, Atlanta, GA 30332-0405, USA; steven.liang@me.gatech.edu

* Correspondence: yuecaixu@hrbust.edu.cn; Tel.: +1-884-693-9745

Received: 19 September 2018; Accepted: 14 November 2018; Published: 9 December 2018

Abstract: It has been proved that surface integrity alteration induced by machining process has a profound influence on the performance of a component. As a widely used processing technology, milling technology can process parts of different quality grades according to the processing conditions. The different cutting conditions will directly affect the surface state of the machined parts (surface texture, surface morphology, surface residual stress, etc.) and affect the final performance of the workpiece. Therefore, it is of great significance to reveal the mapping relationship between working conditions, surface integrity, and parts performance in milling process for the rational selection of cutting conditions. The effects of cutting parameters such as cutting speed, feed speed, cutting depth, and tool wear on the machined surface integrity during milling are emphatically reviewed. At the same time, the relationship between the machined surface integrity and the performance of parts is also revealed. Furthermore, problems that exist in the study of surface integrity and workpiece performance in milling process are pointed out and we also suggest that more research should be conducted in this area in future.

Keywords: milling process; part functionality; surface integrity; research progress

---

1. Introduction

In today's competitive manufacturing industry, the ultimate goal of manufacturers is to produce higher-performance products at lower cost and in less time. At the end of the machining process chain, thermomechanical coupling load has significant impact on workpiece surface property with a direct link to functionality. To get a reliable machined component with high fatigue strength, high wear resistance, and high dimensional accuracy are the goals of the machining process. It is necessary to evaluate the effects of machining parameters, tool parameters, etc. on machined surface performance capabilities.

The specified term surface integrity can be used to evaluate the machined surface properties after manufacturing operations. There are mainly some aspects to describe surface integrity: topography, metallurgy characteristics and residual stress. The topography is made up of surface roughness, waviness and flaws. The metallurgy characteristics include grain size, plastic deformation, microhardness, phase transformation, recrystallization, etc. [1].

The door to the study of machined surface integrity is opened by a review article written by Field et al. [2]. The results show that surface integrity is the intrinsic property of surface hardening conditions that are produced after processing [3]. Then, considerable research on machined surface integrity was carried out [4–8]. The CIRP (The International Academy for Production Engineering) published its keynote paper "Capability Profile of Hard Cutting and Grinding Processes" in 2005 to guide the research of surface integrity; subsequently, a collaborative working group on surface integrity and functional performance of components was established in 2008 [9].

The ultimate research goal of surface integrity is to get the component with high performance capacity. In the cutting process, the finishing pass will define the thermos and mechanical state of the machined surface. Surface integrity has a significant impact on several relevant characteristics of the final functionality of the component, such as dimensional accuracy, friction coefficient, wear and thermal resistance, and fatigue behavior corrosion, as shown in Figure 1. For example, the finished component may lose efficacy for many reasons, such as changes in dimensions due to wear or plastic deformation, deterioration of the surface finish, and cracking or breakage [10]. Therefore, it is important to reveal the effect of the manufacturing process on finish part functionality.

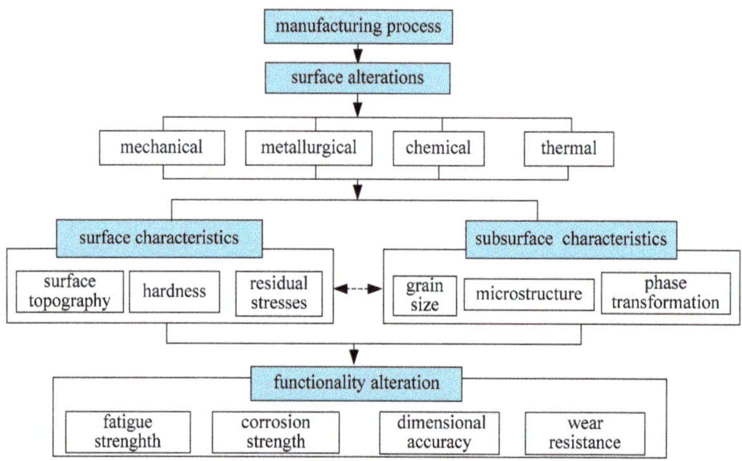

Figure 1. The effect of surface integrity on part functionality.

The definition of surface function differs depending on the operating performance of the surface. For example, when translational surfaces are used as surfaces for the bearing inner and outer rings, tribological functionality will become dominant. However, for the applications of dynamic loading and cyclical stresses, fatigue characteristics may be a prime consideration in determining component failure [11]. Therefore, in order to get the influence mechanism of cutting parameters on parts performance, it is important to study the intrinsic relationship between the factors of surface integrity and parts performance [6].

Milling has the advantages of high production efficiency and wide processing range, and it is widely used in key fields such as aerospace, mold, automobile, and parts manufacturing [12]. To reveal the topic more clearly, the surface integrity and functionality of the part after the milling process are focused on in this paper. The effects of the manufacture process on surface integrity are discussed. Through the analysis research and the finite element modeling method, the surface integrity can be better understood. Then, a review of the state-of-art research on finished component functionality affected by surface integrity, such as surface topography, surface metallography, and residual stress, is presented. In this review, the component functionality mainly includes fatigue strength and wear resistance.

2. Part Functionality as Effected by Surface Topography

2.1. Machined Surface Topography Characteristic

The surface topography of any manufacturing process is a critical index. It is mainly affected by geometric properties of the machining system for milling process. Choosing suitable cutting parameters for given tools, workpiece materials, and machine tools is an important step to get reasonable surface topography. Because the two-dimensional parameters cannot properly represent the surface features of milling, more and more attention is being paid to three-dimensional parameters today.

Considerable research has been performed to get suitable surface topography. Zhang et al. [13] provided in-depth characterization and analysis of the three-dimensional topography of feed direction and cross-feed direction in the hard milling process of AISI H13 steel (AISI is an abbreviation of American Iron and Steel Institute), as shown in the Figure 2. The results show that better surface morphology can be obtained by using higher cutting speed, lower feed speed, and lower cutting depth. Wang et al. [14] studied the surface morphology and surface roughness in the milling process of AlMn1Cu. Due to the high ductility of the AlMn1Cu material, the material flows plastically along the side-cutting edge and meanwhile is extruded by the side-cutting edge to cause the material to accumulate on the machined surface. With the increase of cutting depth and feed per tooth, the plastic flow of the material along the cutting edge is strengthened, resulting in increased surface roughness. However, with the increase of cutting speed, the plastic flow of the material is weakened along the secondary cutting edge, resulting in that the surface roughness value decreases. Ghani et al. [15] found that high cutting speed, low feed rate, and low cutting depth can be used to obtain ideal surface quality in the semifinishing and finishing process of AISI H13 steel.

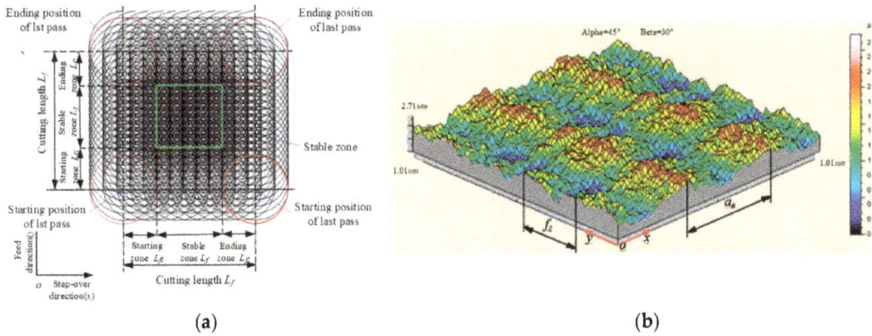

(a)   (b)

Figure 2. Surface topography of milling surface [13]. (**a**) Surface texture zones in milling and (**b**) surface topography by milling process.

Some scholars carried out the research of prediction of surface topography for milling process. As established by experimental tests, the actual roughness values obtained usually deviates from theoretical ones. There are several reasons accounting for this, such as runout of the milling cutter tooth tips, rounding of the cutting edge, and irregularities of cutting edge [16].

In order to design a machining process for blades in turbine engines without prior experiment, Denkena et al. [17] developed a model to predict the surface topography. The effect of stochastic topography on the flow losses was investigated also in his research. The machined surface was studied by combining the dynamic morphology with the random topography based on empirical data. The results show that the random morphology has a great influence on the flow loss and therefore cannot be ignored. Zhang et al. [18] proposed a milling topography simulation model considering tool wear. The experimental results of cutting under plane and cylindrical surfaces are in good agreement with the model prediction, which proves the correctness of the model. Irene et al. [19] established a numerical model to predict the surface topography of milling process based on the contact relationship between tool and workpiece.

Gao et al. presented an analytical model that considered the effects of tool geometry, cutting conditions, and plastic flow measurements to predict surface topography [20]. Arizmendi et al. established a surface topography prediction model considering tool vibration [21]. Lazoglu et al. [22] established an analysis model for the surface topography of five-axis milling process considering cutting parameters, number of cutting edges, and cutter runout.

The cutting trajectory is another factor that should not be neglected in the prediction model of surface topography. Based on the tool machining paths and the trajectory equation of the cutting edge relative to the workpiece, Zhang et al. [23] developed an iterative algorithm for numerical simulation of machining surface topography in multiaxis milling. Gao et al. [24] proposed a new method to predict the machined surface topography with relative cutting edge trajectory equation. Tan et al. explored the effects of different tool paths on the surface morphology, and the results are shown in Figure 3. By comparing Sa and Rt values, it found that vertical upward is the optimal cutting path [25]. Chen et al. [26] studied the effect of different angle combinations on the surface topography in multiaxis milling. Better surface roughness could be achieved when rotation angles are 0 (positive lead), 60 (combination of positive tilt and positive lead), 90 (positive tilt), and 330 (combination of negative tilt and positive lead). Lu et al. [27] studied the micrometer milling mechanism of Inconel 718 and concluded that the status of the machined surface is determined by the profile of the cutter, the cutting path of the cutter and the flexible deformation of the cutter. Therefore, the author comprehensively considered these three factors to build three-dimensional surface topography and a surface roughness prediction model. Compared with the experimental results, the maximum relative error of the model is 10.9% and the average relative error is 6.8%. The results can be used as a reference for the prediction of surface milling.

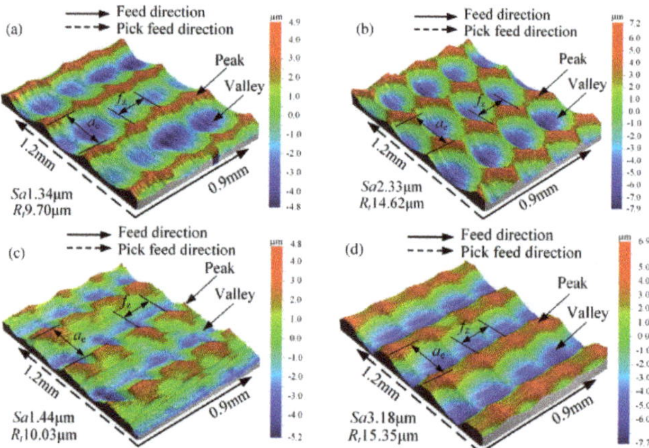

Figure 3. Effect of different cutter path orientations on surface topography [25]: (**a**) Vertical upward; (**b**) vertical downward; (**c**) horizontal upward; and (**d**) horizontal vertical downward.

By combining boundary intersection and mean squared error method, Brito developed the prediction on surface roughness in AISI 1045 steel end milling process. The results showed that the achieved optimum lessens the sensitivity to the variability transmitted [28]. Tangjitsitcharoen et al. used the dynamic milling force ratio to predict the surface roughness of the ball milling process. The model was verified by experiments. The results show that the model can predict the average surface roughness accuracy up to 92.82% [29]. Also, some other research works were developed to get reasonable surface texture [30,31]. Toh et al. analyzed the surface texture generation mechanism under different tool paths by using the method of surface topography analysis, and obtained the best surface texture when the workpiece was inclined at 75° for high speed milling and then determined the optimized machining path [32].

Meanwhile, there may be defect appearing on the machined surface in milling process. Common surface defects are feed mark, tearing surface, and burr formation. Feed marks are produced by the combined effect of tool rotation caused by cutting speed and tool movement caused by feed speed [33]. Feed marks on machined surfaces become more pronounced with increasing cutting speeds and tool wear [34]. Tearing surfaces usually include smeared material, surface microvoids, scratches, and groove and pitting corrosion. The typical tearing surfaces is showed in Figure 4. Damage on the machined surface in hard milling process of FGH95 PM superalloy was investigated by Du and Liu [35]. Their research results showed that several defects appeared on the machined surface at higher cutting speeds. The results are significant for the prediction of component service life.

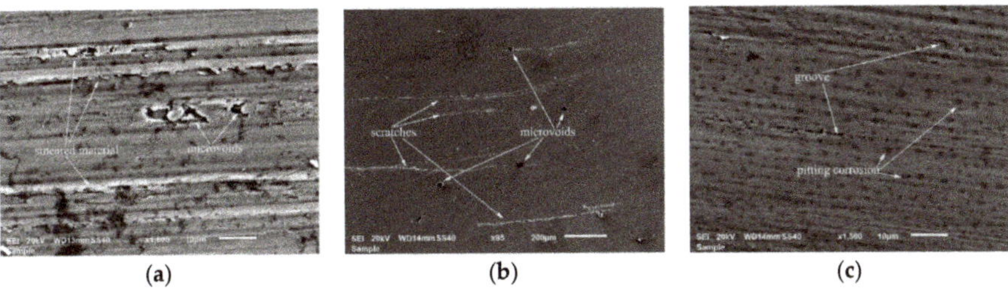

Figure 4. Three kinds of tearing surface [34]. (**a**) Microvoids and scratches; (**b**) microvoids and scratches; and (**c**) groove and pitting corrosion.

When the tool moves along the feed direction, the workpiece material will flow laterally under the combined extrusion action of the minor flank face and the machined surface causing smearing. The surface microvoids are formed from the carbide particles of the workpiece material, the cutting off of the cutting tools, and the deposition of the chip. The surface microvoids affect the mechanical properties of the workpiece, so the surface microvoids should be avoided in the key components [36].

Also, burr defects are often observed in the area where the tool exits the cutting zone, and the formation of burr is easy to cause stress concentration and reduce the fatigue life of the components. Generally, the materials with low thermal conductivity and Young's modulus are more likely to form burrs on the cutting surface [37]. Under normal circumstances, the burr decreases with the increase of cutting speed, and increases with the increase of tool wear [34].

Generally, cutting parameters with high cutting speed, low feed rate, and low cutting depth is recommend for a good surface finish for milling process.

2.2. Fatigue Strength and Wear Resistance as Effected by Surface Topography

Milling process is generally applied to conduct finishing machining of sculptured surfaces, therefore the machined component have very demanding specifications in surface topography for its performance. The smoothest surfaces are desired in most milling processes, especially when the fatigue life of the part being machined is high [38]. However, in some biomedical fields, it is desirable to have rough surface morphology [39]. The surface structure is responsible for the mechanical functionality of the component. Sometimes, even if the surface dimension and surface finish of a component are well within the tolerance, there remains the possibility of lack of surface quality for a milled surface. The reason is surface topography influences not only the mechanical and physical properties of contacting parts, but also optical and coating properties of some no contacting components.

Because of the stress concentration caused by pits and groove, the characteristics of milling morphology have important influence on its performance. Generally, rougher surfaces are expected to encourage fatigue crack initiation. It is suggested that parameters such as Rt and Rz are more appropriate than Ra in respect of fatigue strength, as they equate with adverse component surface features [38]. Arola et al. [40] used surface roughness to estimate the effective stress concentration coefficient of high strength low alloy steel workpiece. It was found that the fatigue strength decreased with the increase of surface roughness at low stress level. Moussaoui et al. studied the factors that affect the fatigue life of titanium alloys. They found that the surface roughness does not affect the fatigue life of the workpiece [41]. Guo et al. found that the surface roughness has a slightly influence on four-point bending fatigue life for end milling process of AISI H13 [42]. Li et al. [43] found that surface roughness has little influence on fatigue life due to the small ratio between the ten-point surface height and the bottom curvature radius of surface gaps. Wang et al. [44] found that high surface roughness leads to a high stress concentration coefficient, which reduces the fatigue life of the workpiece.

The machined surface morphology affects the fatigue performance of the final assembly, especially when the crack initiation life is noteworthy [45,46]. Therefore, many works have been done to explore the effect of surface topography on part performance. The traditional two-dimensional surface roughness does not adequately characterize the effect of surface properties on the fatigue performance, while the three-dimensional topography can provide a more accurate correlation with fatigue behavior [47–49].

Performance and reliability of engineering components can be enhanced by selecting appropriate 3D topographic characteristics [50]. Novovicetc et al. investigated the effected of surface and subsurface condition on the fatigue life of Ti alloy workpiece. Their research results indicated that the surface topography, in particular texture direction, showed a strong correlation with the fatigue life of workpiece [51]. Piska et al. [52] studied the effect of progressive milling process on the surface morphology and fatigue properties of 7475-T7351 high-strength aluminum alloy. The results show that the surface topography is not the right factor affecting the fatigue properties of aluminum alloys.

When the roughness level is the same, the milling surface has a higher fatigue limit than the grinding surface. The reason is that the surface grooves produced by milling process are more random [53]. Huang et al. [54] studied the effect of different tool paths on the fatigue life of AISI H13 high speed milling process. It was noted that different tool paths lead to the difference of microscopic stress concentration caused by the orientation morphology, which affects the fatigue performance of the workpiece.

The surface integrity after machining directly affects the life and reliability of the workpiece, so it is important to study the effect of different surface integrity on the friction and wear properties of the workpiece surface to improve the service life of the workpiece. Sedlaek et al. [55] studied the relationship between surface roughness parameters and friction and wear properties of mold surface. The results show that the friction coefficient is inversely proportional to the surface roughness under the condition of dry friction, and the change trend is opposite in the lubrication conditions. Menezes et al. [56] studied the effect of surface texture on friction coefficient and adhesion wear in friction pair. The results show that the friction coefficient and adhesion wear are mainly influenced by the surface texture, which is independent of the surface roughness. Magri et al. [57] studied the relationship between surface morphology and wear resistance during die milling. They found that the best tribological performance was that composed of microcavities generated by similar and high values of fz and ae. The surface topography under four cutting conditions and the die wear results is shown in Figure 5.

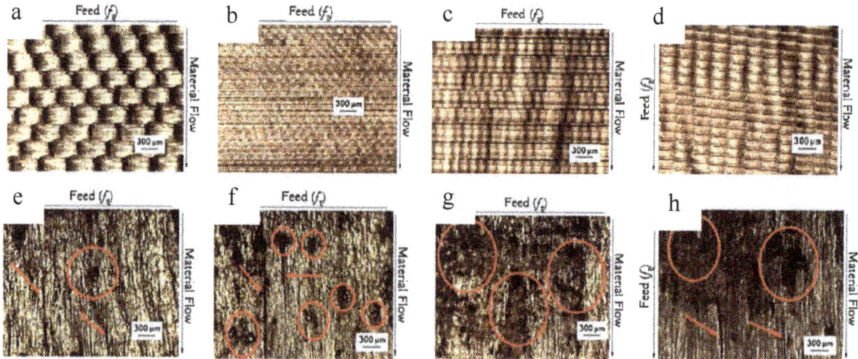

Figure 5. Four types of surfaces topography and die flash land after forging [57]. Profiles of the die flash land surfaces under (**a**) condition 1, (**b**) condition 2, (**c**) condition 3, and (**d**) condition 4; Die flash land after forging under (**e**) condition 1, (**f**) condition 2, (**g**) condition 3, and (**h**) condition 4.

After milling processes, surface topography has a direct impact on part functional performance, especially in respect of fatigue life and wear resistance. The reason is the surface topography has close relationship with surface frictional characteristic and stress distribution of milled surface. To get a good performance part, the surface topography should be selected according to the application condition of workpiece.

3. Part Functionality as Effected by Surface Metallurgy

3.1. Machined Surface Metallurgy Characteristic

In milling of high hardness material or difficult–to-machine metal material, high stress, strain rate, and temperature will have a severe impact on the machined surface. The microscale and nanometer scale existing on the machined surface will change under the interaction of large strain, high strain rate and high temperature [58,59]. The property changes include microstructure change, plastic deformation. Many research works have been done on process mechanics and surface integrity due to the complex coupling between phase transformations and loading in milling process [60]. Generally, the depth of microstructural alteration has been observed to increase when the cutting speed, feed rate and tool wear are increased.

For milling process, Elbestawi et al. investigated the microstructural alterations in high speed milling of hardened AISI H13 using PCBN (Polycrystalline Cubic Boron Nitride) ball-nose end tools. They found that the formation of phase transformation layer was dependent on tool edge preparation and tool wear [61].

Both thermal and mechanical effects attribute to the plastic deformations significantly. Mechanical effects play a major role in the hardening of materials, while thermal effects play a major role in the softening of materials. Moreover, the rubbing effect of flank face on machined surface play an important role in its generation. That is why the tool wear increases from initial condition to its life value, the changed microstructure turns to be deeper. The effects of cutting speed and tool wear on the surface microstructure were investigated when milling titanium alloy Ti-1023 [34]. When VB = 0 (Tool wear value), the cutting speed has no effect on the phase transition or deformation of the machined surface, as shown in Figure 6a. Tool Wear has a significant effect on the plastic deformation and the depth of microstructure of the machined surface, as shown in Figure 6b.

The reasonable selection of cutting parameters in the finishing process does not cause the change of the structure type of the machined surface. The reason is the contact area between the cutting tool and the machined surface is so smaller that the maximal temperature of the machined surface would be lower than the austenitizing temperature of workpiece material [62]. The similar results were found by Li and Zhao [63] and Devillez et al. [64].

Extensive experimental work has been conducted to get an ideal surface metallurgical for machining process. However, indispensable hardware is needed for this method. So the modeling for machined surface microstructure has attracted wide attention [65–69]. Unfortunately, few research works focused on the milling process are found.

After high-speed milling of Ti-6Al-4V and Ti-834, the microstructural subsurface damage in the form of intense slip bands was identified by Thomas et al. Due to a reduction in fatigue crack initiation resistance, the microstructural subsurface damage could degrade the in-service properties of workpiece [70]. Shyha et al. [71] studied the influence law of cutting fluid supply system on metallurgical characteristics, and concluded that cutting fluid had little influence on the microstructure and deformation layer of cutting subsurface; cutting speed was a key factor affecting microstructure. Li et al. [72] carried out an experimental study on hard milling of AISI H13 steel, and the results showed that the nanohardness and plastic deformation depth of the machined surface increased with the increase of the grinding radius of the cutting edge. Liu et al. [73] conducted an experimental study on the AA7150-T651 aluminum alloy. It was found that the severe shear strain caused by the mutual friction between the workpiece and the cutter resulted in a high deformation layer near the surface area.

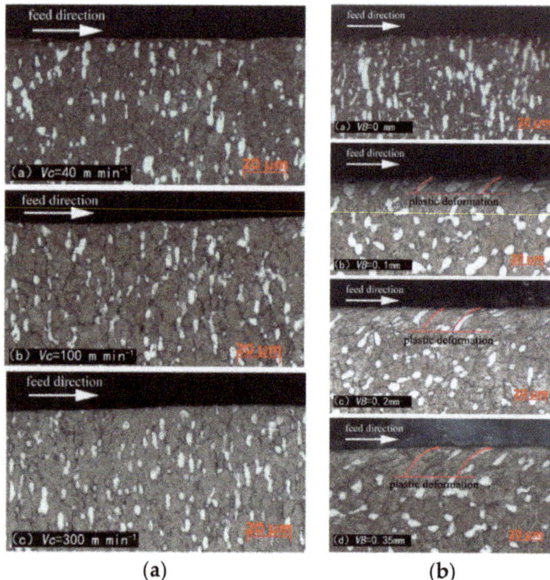

Figure 6. Effects of cutting speed and tool wear on subsurface microstructure [34]. (**a**) VB = 0 m and (**b**) V_c = 60 m min^{-1}, f_z = 0.08 mm/tooth, a_p = 1 mm.

Generally, if the temperature generating on the milled surface is higher than austenitizing temperature of workpiece material, then there will be phase transformations layer appearing on the milled surface. Combining with mechanic load in milling process, the thermal load will make the surface and subsurface metallurgy change. Additionally, the effect of tool should not be neglected. Because the cutting tool will cut-out and the tool will cut-in during the milling process, the thermal and mechanic loads applied to the workpiece do not remain the same. So, the relationship between surface metallurgy characteristic and cutting parameters should be investigated future.

3.2. Wear Resistance and Corrosion Resistance as Effected by Surface Metallurgy

The microstructural alterations in the material cause the surface layer to exhibit different material behavior, and these alterations include phase transformations and plastic deformations. After the workpiece is machined, the behavior and property of surface is different with the interior of the bulk material. So, the subsurface microstructure has a crucial impact on the performance of the final part. There are controversial standpoints about whether the formation of phase transformations is beneficial to the application. It is more brittle than bulk material, so the appearance of the white layer usually worsens the product's service life [74]; therefore it is important to prevent its occurrence or at least predict how it would affect the final product.

There is similar pattern of higher hardness on the milled surface because of microstructural alteration. In the production process, the effect of hardening layer can be eliminated when the cutting depth is greater than the hardening layer of the workpiece, but it is difficult to realize [75]. Minimizing or eliminating the phase transformations layer would improve the machined surface quality and fatigue strength. Microhardness is a comprehensive index characterizing the microstructure of surface materials, which can be used to characterize the effect of microstructure on fatigue performance. Related research shows that the increase of microhardness in a certain range can improve the fatigue life of workpieces [76,77].

Fellah et al. [78] found that crystallite and grain size play a controlling role in friction coefficient and wear rate. The smaller the grain size, the higher the wear resistance. Zhao et al. [79] studied the friction characteristics of the machined surface of titanium alloy and found that the wear resistance of the workpiece material increases gradually with the increase of hardening degree and grain refinement degree of the machined surface. Huang et al. [80] studied the friction and wear behavior of milling AISI D2 steel. It is found that the subsurface grain deformation induced by machining is helpful to improve the wear resistance of the workpiece. At the same time, it is pointed out that the depth of deformation zone and grain boundary inclination angle can be used as the evaluation index of wear resistance to some extent.

The corrosion resistance of 7050-T7451 aluminum alloy processed by high speed milling was studied. It is found that the surface corrosion damage of the workpiece is determined by the interface energy between the grains and the degree of hardening [81].

Additionally, Kim et.al. used two different methods to estimate the effect of cooling methods on hot forging die service life against plastic deformation and abrasive wear. They found that the die service life depended on abrasive wear, rather than the plastic deformation of the die, for a specific cooling method [82]. To enhance the functional properties of the machined surface, roller burnishing is an effective approach as it changes the microstructure of surface and subsurface [6].

The alterations of surface metallurgy have an important impact on part functional performance. Also the effects of surface metallurgy on part performance are complicated in different application conditions. To get better performance of milled workpiece, the effect of phase transformations and plastic deformations in surface metallurgy should be considered comprehensively.

4. Part Functionality as Effected by Surface Residual Stress

4.1. Residual Stress Characteristic on Machined Surface

Residual stresses are stresses that remain in a solid body after the external loading (mechanical and thermal) has been eliminated [83]. At the same time, the microstructure changes of machined surface can also cause residual stress [84]. Some scholars have studied the basic principle of residual stress caused by machining [33,85,86]. As far as 1982, Brinksmeier et al. have conducted such research. They reported the causes of residual stresses in machining process [87]. The final state of residual stress on the component has close relationships with other surface integrity factors, such as surface topography, subsurface microstructure, and topological states of a machined surface.

Considerable research on residual stress has been done for milling process. Titanium alloy is widely used in aerospace industry because of its high mechanical strength, chemical resistance, and thermal conductivity, so it is very important to control the distribution of residual stress. Sun et al. studied the surface integrity in the process of Ti-6Al-4V milling by the experimental method. It was found that the compressive residual normal stresses in the cutting and feed direction increase with the increase of cutting speed. Meanwhile, it is found that the compressive residual normal stress in feed direction is larger than that in cutting direction [88]. Rao et al. studied the milling of Ti-6-Al-4V titanium alloy and concluded that the compressive residual stress increased with the increase of feed and the cutting speed had no effect on the compressive residual stress [89]. For milling process of Inconel 718 with carbide K20, the increase of milling speed will enhance the tensile stress at the surface and compressive stress beneath the machined surface.

Aspinwall et al. used experimental methods to study the effect of tool positioning and workpiece angle on machining performance during milling Inconel 718. Compressive residual stress is generated in horizontal upward cutting and tensile residual stress is generated in horizontal downward cutting. The reason is that horizontal upward has a relatively low cutting speed, so the local temperature of the machined surface is lower. The influence of tool wear and cutting direction on residual stress is shown in Figure 7 [90]. Jiang et al. investigated the effects of tool diameters on the residual stress in the milling process of a thin-walled part. They found that the distribution of residual stress was more uniform as the tool diameter increased [91]. For milling process of nickel alloy and titanium alloy, many research results indicated that increasing cutting speed will bring about the tensile residual stresses tend to become more compressive [92]. Axinte et al. studied the residual stresses in the feed direction during milling AISI H13. It is found that cutting speed and feed per tooth are important factors affecting the residual stress on the surface [93]. When AISI H13 steel is hard milled by coated cutting tools, the investigated result of in-depth residual stress distribution showed that microstructural changes deeply affect the residual stress and that they have to be accurately taken into account during the process design [94]. Wang et al. [95] studied the effects of cutting parameters, cutting fluid, and spindle angle on residual stress during milling of Inconel 718 alloy. The results show that the cutting depth and cutting speed have great influence on the distribution of residual stress, and the residual stress in the tensile direction increases gradually with the increase of the spindle angle.

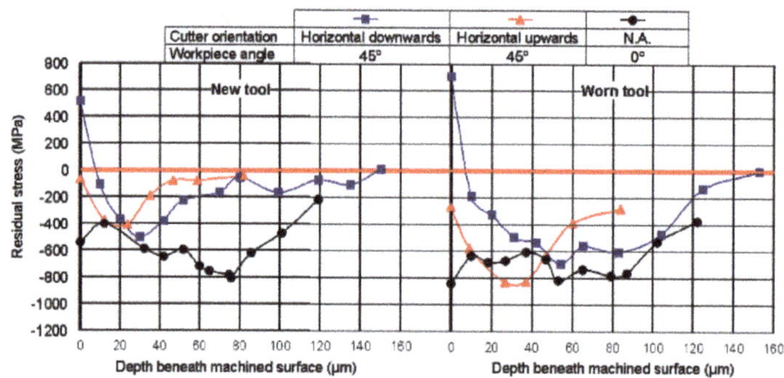

Figure 7. Effect of different cutter orientations/workpiece tilt angle on residual stress [90].

Although the experimental method can directly reflect the relationship between the distribution of residual stress and cutting conditions, it has higher cost and longer period. Therefore, it is urgent to develop a predictive model of residual stress distribution in milling process. Jiang et al. [96] used finite element simulation to study the distribution of residual stress in high-speed peripheral milling, and found that the cutting thickness had a significant impact on the tangential residual stress. Li et al. [97] used the finite element method to study the effect of cutting depth on the residual stress in milling, and pointed out that the residual stress can be reduced by optimizing the cutting depth of thin wall parts. Liang et al. [98] established an exponential decay function considering the flank wear, tool inclination, and depth of cut to predict the compressive residual stresses distribution of the milled TC17 alloy (An alpha-beta titanium alloy). Fergani et al. [99] used Neumann–Duhamel principle to establish the regeneration prediction model of residual stress in multipass machining under thermomechanical loading. Considering the three-dimensional instantaneous contact state between the milling cutter and the workpiece, Wan et al. [100] established the theoretical model for predicting the residual stress in the milling process. At the same time, it was observed that the thermal load had a relatively weak effect on the residual stress.

Residual stress is a crucial factor to evaluate surface integrity of the milled workpiece. In milling process, distribution of residual stresses is marginally affected by the cutting speed and tool wear state. Also, its residual distribution has close relationship with cutting strategy in milling process. To get reasonable distribution of residual stresses, the cutting parameters should be optimized with consideration of tool parameters.

4.2. Fatigue Resistance as Effected by Surface Residual Stress

The residual stress has an important effect on the mechanical properties of the workpiece. Residual stresses directly influence the deformation of workpiece, such as static and dynamic strength and chimerical and electrical properties, as shown in Figure 8. It is well known that surface finish and residual stress can significantly affect the antidestructive ability of the component under high cyclic fatigue load.

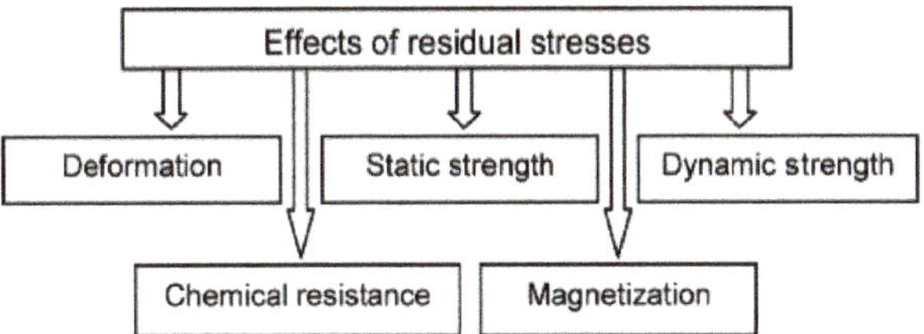

Figure 8. Effect of residual stress on workpiece performance.

Compressive residual stress is usually advantageous to the fatigue life of machined parts, while residual tensile stress is the opposite. The tensile residual stress will enlarge or contribute the extension of microcrack. When the crack increases to a certain extent, it will cause workpiece failure. So, it is necessary to remove tensile residual stresses occurring during machining process. In view of the high speed milling process of Ti-10V-2Fe-3Al, Yao et al. [101] found that the fatigue life of the workpiece is more sensitive to residual stress than surface roughness. Similar results were found by Moussaoui et al. [41].

Influence of milling process on the fatigue life of Ti6-Al-4V was investigated by Moussaoui et al. They found that residual stress has a more preponderant influence on fatigue life than geometric and metallurgical parameters [102]. Huang et al. [55] studied the effect of different tool paths on the fatigue life of AISI H13 high-speed milling process. The results show that the different tool paths lead to the difference of effective residual stress, which results in different fatigue performance of the workpiece. Meanwhile, it is found that the influence of effective residual stress on the fatigue life of the workpiece is greater than that of the orientational surface topography. Yang et al. [103] studied the effect of residual stress on the fatigue life of workpieces during milling of Ti-6Al-4V. The results show that increasing the compressive residual stress can effectively improve the fatigue performance of the workpiece. However, when the workpiece surface material produces plastic deformation, the effect of residual stress will disappear.

Residual stress is a crucial factor to evaluate surface integrity of the milled workpiece. In milling process, distribution of residual stresses is marginally affected by the cutting speed and tool wear state. Also, its residual distribution has close relationship with cutting strategy in milling process. To get reasonable distribution of residual stresses, the cutting parameters should be optimized with consideration of tool parameters. If the distribution of residual stresses is not reasonable, it can be modified due by mechanical loads (static or cyclic) or thermal exposure [104].

The influence of residual stress and microstructure on the fatigue life of the workpiece was studied by F. Ghanem et al. By comparing electro-discharge machining (EDM) with the milling process, they found that the surfaces prepared by EDM showed a tensile residual stress at the surface. The milled surfaces showed a near-surface compressive residual stress, which is favorable for fatigue crack resistance, and the comparison result is shown in Figure 9 [105]. Three-point bending fatigue tests of the notched specimens revealed a loss of 35% in fatigue endurance in the case of EDM.

Tool wear is an unavoidable and complicated phenomenon occurring in machining process, which has a direct impact on residual stress on milled surface. For milling process, many workpiece materials contain carbide particles in their structure. As the cutting tool wears, some carbide particles in workpiece are removed from the machined surface sometimes, then there will be direct effect on the surface quality of the machined surface. Generally, tool flank wear was found to have the most effect on distribution of residual stress in high speed milling process.

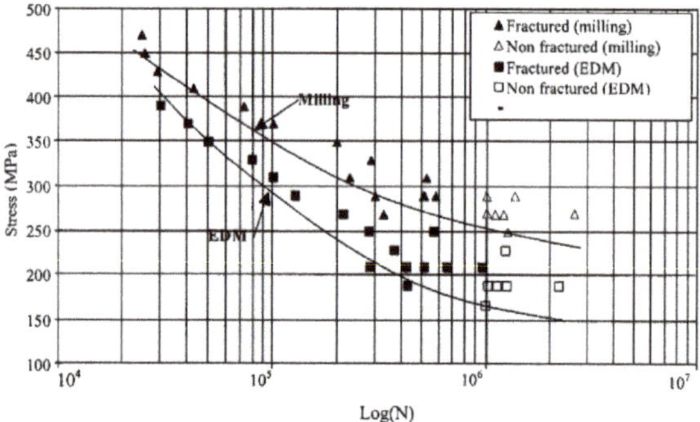

Figure 9. Comparison effects of residual stress on fatigue life between electro-discharge machining and milling [105].

Considerable research has been performed to get the effect of tool wear on part functionality. The effects of tool wear on surface integrity and fatigue life during milling was studied by Guo et al. They found that surface roughness is generally higher in the step-over direction than the feed direction and that tool wear does not necessarily affect the fatigue life within a certain range (0.2 mm) [106]. Also, fundamental relationship between tool wear and fatigue was related by them. They found that compared with the sharp tool, the worn tool up to flank wear (0.12 mm) only slightly reduces the average life from 1.23 × 106 to 1.16 × 106 cycles. Rougher surfaces are expected to encourage fatigue crack initiation. While Guo found that the surface roughness also has a slightly influence on four-point bending fatigue life for end milling of AISI H13 [42], and the effect of tool wear on workpiece fatigue life is shown in Figure 10.

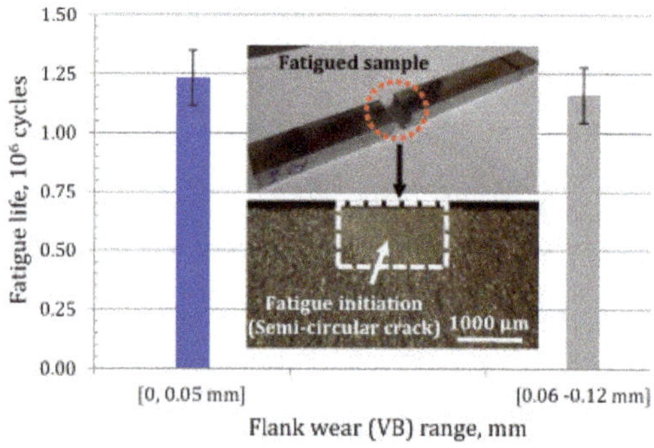

Figure 10. Fatigue life vs. tool wear and fatigue fracture pattern [42].

The residual stress on milled surface will directly influence the generation and extension of crack on surface, and that will have a direct impact on fatigue life of workpiece. Generally, compressive residual stress contributes to the improvement of fatigue life. Tool wear has a direct impact on the residual stress of the milled surface, so tool wear is another important factor that influences the fatigue life of component in milling process.

5. Conclusions and Outlook

The optimization and control of machined surface integrity is the key technology to ensure the functional performance and service life of parts. The effects of cutting parameters such as cutting speed, feed speed, cutting depth, and tool wear on the machined surface integrity during milling are emphatically reviewed. At the same time, the research progress of the relationship between the machined surface integrity and the component use performance is also revealed. Researchers have made extensive research on the influence of machined surface integrity on fatigue properties and obtained a series of research results. However, there are still some urgent problems to be solved.

1. The effects of surface integrity on performance were mostly determined by experiments and could not be revealed effectively at the mechanism level.
2. No uniform standard has been formed to assess the effect of surface integrity parameters on fatigue performance, which limits the optimization of process parameters and the improvement of service performance of workpieces.
3. The local performance change, design, and service performance state of the parts could not be organically combined to form an interconnected mode.
4. Combined with the current research status and engineering application requirements, the following aspects can be emphatically studied in the future.
5. To strengthen theoretical research on fatigue performance optimization. The existing optimization models assume that the surface integrity evaluation indices (surface morphology, residual stress, microhardness, and metamorphic layer thickness) have the same effect on fatigue performance, so the weight of the coefficient of the optimization model is consistent. In order to make the optimization result more accurate, the corresponding mathematical analysis method can be considered to distinguish the weight of various indexes.
6. To strengthen the research on modeling technology of machined surface integrity and workpiece performance. The analytic model of residual stress of workpiece surface during milling is still a difficult problem. The reason is that the change of chip thickness on cutting edge is more complicated, which results in the nonlinear gradient distribution of cutting force and cutting temperature in machining process. Only some physical factors (vibration, tool/workpiece deformation, and cutting force) are considered in the analytical model of surface topography, while tool wear and the characteristics of interrupted cutting are not taken into account, so there is still a large error in the simulation results.
7. Development of new milling processes. The reasonable selection of process parameters directly affects the machined surface integrity and controls the workpiece performance. How to develop a new milling process based on different machining systems, processing environment, workpiece materials, and performance needs to be further explored.
8. The relationship between the design-manufacturing-service performance of parts is explored. A complete interconnected mode of design, manufacturing, and service performance evaluation of parts was established.
9. The theoretical analysis and experimental research on the response relationship between high temperature fatigue, vibration fatigue, corrosion fatigue performance, and surface integrity of processing under environmental conditions are further improved.

Author Contributions: The idea of this project was conceived by C.Y. C.Y and H.G. consulted to relevant high-level papers on aspects of changes in surface integrity caused by changes in the milling process, and wrote this project. X.L. and S.Y.L. reviewed this project and proposed constructive guidance to make the article more complete.

Funding: This research was funded by the National Natural Science Foundation of China (Grant No. 51575147) and the Science Funds for the Young Innovative Talents of HUST (No. 201507), and the International Cooperation and Exchanges NSFC (No.51720105009).

References

1. Field, M.; Kahles, J.F.; Cammett, J. Review of measuring methods for surface integrity. *CIRP* **1972**, *21*, 219–238.
2. Field, M.; Kahles, J.F. *Review of Surface Integrity of Machined Components*; Defense Technical Information Center: Fort Belvoir, VA, USA, 1971.

3.   Field, M.; Kahles, J. The surface integrity of machined-and ground high-strength steels. *DIMC Rep.* **1946**, *210*, 54–58.

4.   Yue, C.; Liu, X.; Ma, J.; Liu, Z. Hardening effect on machined surface for precise hard cutting process with consideration of tool wear. *Chin. J. Mech. Eng.* **2014**, *27*, 1249–1256. [CrossRef]

5.   Yue, C.; Liu, X. Research progress of machined surface integrity for high stress steel cutting process. *J. Harbin Univ. Sci. Technol.* **2012**, *16*, 5–10.

6.   Jawahir, I.; Brinksmeier, E.; M'saoubi, R.; Aspinwall, D.; Outeiro, J.; Meyer, D.; Umbrello, D.; Jayal, A. Surface integrity in material removal processes: Recent advances. *CIRP Ann. Manuf. Technol.* **2011**, *60*, 603–626. [CrossRef]

7.   Ulutan, D.; Ozel, T. Machining induced surface integrity in titanium and nickel alloys: A review. *Int. J. Mach. Tool Manuf.* **2011**, *51*, 250–280. [CrossRef]

8.   Davim, J.P. *Surface Integrity in Machining*; Springer: Berlin, Germany, 2010.

9.   Klocke, F.; Brinksmeier, E.; Weinert, K. Capability profile of hard cutting and grinding processes. *CIRP Ann. Manuf. Technol.* **2005**, *54*, 22–45. [CrossRef]

10.   M'Saoubi, R.; Outeiro, J.; Chandrasekaran, H.; Dillon, O.; Jawahir, I. A review of surface integrity in machining and its impact on functional performance and life of machined products. *Int. J. Sustain. Manuf.* **2008**, *1*, 203–236. [CrossRef]

11.   Aris, N.; Cheng, K. Characterization of the surface functionality on precision machined engineering surfaces. *Int. J. Adv. Manuf. Technol.* **2008**, *38*, 402–409. [CrossRef]

12.   Budak, E. Analytical models for high performance milling. Part I: Cutting forces, structural deformations and tolerance integrity. *Int. J. Mach. Tool Manuf.* **2006**, *46*, 1478–1488. [CrossRef]

13.   Zhang, S.; Guo, Y. Taguchi method based process space for optimal surface topography by finish hard milling. *J. Manuf. Sci. Eng. ASME* **2009**, *131*, 051003. [CrossRef]

14.   Wang, Z.; Yuan, J.; Yin, Z.; Hu, X. Surface topography and roughness of high-speed milled AlMn1Cu. *Chin. J. Mech. Eng.* **2016**, *29*, 1200–1207. [CrossRef]

15.   Ghani, J.A.; Choudhury, I.; Hassan, H. Application of Taguchi method in the optimization of end milling parameters. *J. Mater. Process. Technol.* **2004**, *145*, 84–92. [CrossRef]

16.   Li, Z.L.; Zhu, L.M. Envelope surface modeling and tool path optimization for five-axis flank milling considering cutter runout. *J. Manuf. Sci. Eng. ASME* **2014**, *136*, 041021. [CrossRef]

17.   Denkena, B.; Böß, V.; Nespor, D.; Gilge, P.; Hohenstein, S.; Seume, J. Prediction of the 3D Surface Topography after Ball End Milling and its Influence on Aerodynamics. *Procedia CIRP* **2015**, *31*, 221–227. [CrossRef]

18.   Zhang, C.; Guo, S.; Zhang, H.; Zhou, L. Modeling and predicting for surface topography considering tool wear in milling process. *Int. J. Adv. Manuf. Technol.* **2013**, *68*, 2849–2860. [CrossRef]

19.   Irene, B.C.; Joan, V.C.; Alejandro, D.F. Surface topography in ball-end milling processes as a function of feed per tooth and radial depth of cut. *Int. J. Mach. Tool Manuf.* **2012**, *53*, 151–159.

20.   Gao, Y.; Sun, R.; Chen, Y.; Leopold, J. Analysis of chip morphology and surface topography in modulation assisted machining. *Int. J. Mech. Sci.* **2016**, *111*, 88–100. [CrossRef]

21.   Arizmendi, M.; Campa, F.J.; Fernández, J.; Gil, A.; Bilbao, E.; Veiga, F.; Lamikiz, A. Model for surface topography prediction in peripheral milling considering tool vibration. *CIRP Ann. Manuf. Technol.* **2009**, *58*, 93–96. [CrossRef]

22.   Lazoglu, I. 3D surface topography analysis in 5-axis ball-end milling. *CIRP Ann. Manuf. Technol.* **2017**, *66*, 133–136.

23.   Zhang, W.H.; Tan, G.; Wan, M.; Gao, T. A new algorithm for the numerical simulation of machined surface topography in multiaxis ball-end milling. *J. Manuf. Sci. Eng. ASME* **2008**, *130*, 284. [CrossRef]

24.   Gao, T.; Zhang, W.; Qiu, K.; Wan, M. Numerical simulation of machined surface topography and roughness in milling process. *J. Manuf. Sci. Eng. ASME* **2006**, *128*, 96–103. [CrossRef]

25.   Tan, L.; Yao, C.; Ren, J.; Zhang, D. Effect of cutter path orientations on cutting forces, tool wear, and surface integrity when ball end milling TC17. *Int. J. Adv. Manuf. Technol.* **2016**, *88*, 1–14. [CrossRef]

26.   Chen, X.; Zhao, J.; Dong, Y.; Li, A.; Wang, D. Research on the machined surface integrity under combination of various inclination angles in multi-axis ball end milling. *Proc. Inst. Mech. Eng. B J. Eng.* **2014**, *228*, 31–50. [CrossRef]

27.   Lu, X.; Hu, X.; Jia, Z.; Liu, M.; Gao, S.; Qu, C. Model for the prediction of 3D surface topography and surface roughness in micro-milling Inconel 718. *Int. J. Adv. Manuf. Technol.* **2017**, *94*, 2043–2056. [CrossRef]

28. Brito, T.G.; Paiva, A.P.; Ferreira, J.R.; Balestrassi, P. A normal boundary intersection approach to multiresponse robust optimization of the surface roughness in end milling process with combined arrays. *Precis. Eng.* **2014**, *38*, 628–638. [CrossRef]

29. Tangjitsitcharoen, S.; Thesniyom, P.; Ratanakuakangwan, S. Prediction of surface roughness in ball-end milling process by utilizing dynamic cutting force ratio. *J. Intell. Manuf.* **2017**, *28*, 13–21. [CrossRef]

30. Benardos, P.; Vosniakos, G.-C. Predicting surface roughness in machining: A review. *Int. J. Mach. Tool Manuf.* **2003**, *43*, 833–844. [CrossRef]

31. Benardos, P.; Vosniakos, G.C. Prediction of surface roughness in CNC face milling using neural networks and Taguchi's design of experiments. *Robot. Comput. Integr. Manuf.* **2002**, *18*, 343–354. [CrossRef]

32. Toh, C.K. Surface topography analysis in high speed finish milling inclined hardened steel. *Precis. Eng.* **2004**, *28*, 386–398. [CrossRef]

33. Ginting, A.; Nouari, M. Surface integrity of dry machined titanium alloys. *Int. J. Mach. Tool Manuf.* **2009**, *49*, 325–332. [CrossRef]

34. Yang, H.; Chen, Z.; Zhou, Z. Influence of cutting speed and tool wear on the surface integrity of the titanium alloy Ti-1023 during milling. *Int. J. Adv. Manuf. Technol.* **2015**, *78*, 1113–1126.

35. Du, J.; Liu, Z.; Lv, S. Deformation-phase transformation coupling mechanism of white layer formation in high speed machining of FGH95 Ni-based superalloy. *Appl. Surf. Sci.* **2014**, *292*, 197–203. [CrossRef]

36. Xue, C.; Chen, W.Y. Surface damages generated in hole making of a cast nickel-based alloy. *Mater. Sci. Forum* **2012**, *697*, 57–60.

37. Blotter, L.K.G.P.T. The formation and properties of machining burrs. *J. Eng. Ind.* **1976**, *98*, 66–74.

38. Novovic, D.; Dewes, R.; Aspinwall, D.; Voice, W.; Bowen, P. The effect of machined topography and integrity on fatigue life. *Int. J. Mach. Tool Manuf.* **2004**, *44*, 125–134. [CrossRef]

39. Chrzanowski, W.; Neel, E.A.A.; Armitage, D.A.; Knowles, J. Effect of surface treatment on the bioactivity of nickel–titanium. *Acta Biomater.* **2008**, *4*, 1969–1984. [CrossRef] [PubMed]

40. Arola, D.; Williams, C.L. Estimating the fatigue stress concentration factor of machined surfaces. *Int. J. Fatigue* **2002**, *24*, 923–930. [CrossRef]

41. Moussaoui, K.; Mousseigne, M.; Senatore, J.; Chieragatti, R. The effect of roughness and residual stresses on fatigue life time of an alloy of titanium. *Int. J. Adv. Manuf. Technol.* **2015**, *78*, 557–563. [CrossRef]

42. Li, W.; Guo, Y.; Guo, C. Superior surface integrity by sustainable dry hard milling and impact on fatigue. *CIRP Ann. Manuf. Technol.* **2013**, *62*, 567–570. [CrossRef]

43. Li, X.; Zhao, P.; Niu, Y.; Guan, C. Influence of finish milling parameters on machined surface integrity and fatigue behavior of Ti1023 workpiece. *Int. J. Adv. Manuf. Technol.* **2017**, *91*, 1297–1307. [CrossRef]

44. Wang, X.; Huang, C.; Zou, B.; Liu, G.; Zhu, H. Experimental study of surface integrity and fatigue life in the face milling of inconel 718. *Front. Mech. Eng.* **2018**, *13*, 243–250. [CrossRef]

45. Noll, G.C.; Erickson, G.C. Allowable stresses for steel members of finite life. *Proc. Soc. Experts Stress Anal.* **1948**, *5*, 132–143.

46. McKelvey, S.A.; Fatemi, A. Surface finish effect on fatigue behavior of forged steel. *Int. J. Fatigue* **2012**, *36*, 130–145. [CrossRef]

47. De Chiffre, L.; Christiansen, S.; Skade, S. Advantages and industrial applications of three-dimensional surface roughness analysis. *CIRP Ann. Manuf. Technol.* **1994**, *43*, 473–478. [CrossRef]

48. Waikar, R.A.; Guo, Y.B. A comprehensive characterization of 3D surface topography induced by hard turning versus grinding. *J. Mater. Process. Technol.* **2008**, *197*, 189–199. [CrossRef]

49. Yang, D.; Liu, Z.; Xiao, X.; Xie, F. The Effects of Machining-induced Surface Topography on Fatigue Performance of Titanium Alloy Ti-6Al-4V. *Procedia CIRP* **2018**, *71*, 27–30. [CrossRef]

50. Dong, W.; Sullivan, P.; Stout, K. Comprehensive study of parameters for characterizing three-dimensional surface topography: III: Parameters for characterizing amplitude and some functional properties. *Wear* **1994**, *178*, 29–43. [CrossRef]

51. Novovic, D.; Aspinwall, D.K.; Dewes, R.C.; Bowen, P.; Griffiths, B. The effect of surface and subsurface condition on the fatigue life of Ti–25V–15Cr–2Al–0.2 C% wt alloy. *CIRP Ann. Manuf. Technol.* **2016**, *65*, 523–528. [CrossRef]

52. Piska, M.; Ohnistova, P.; Hornikova, J.; Hervoches, C. A Study of Progressive Milling Technology on Surface Topography and Fatigue Properties of the High Strength Aluminum Alloy 7475-T7351. In Proceedings of the International Conference on New Trends in Fatigue and Fracture, Cancun, Mexico, 25–27 October 2017; Springer: Cham, Switzerland, 2017; pp. 7–17.

53. Taylor, D.; Clancy, O. The fatigue performance of machined surfaces. *Fatigue Fract. Eng. Mater. Struct.* **1991**, *14*, 329–336. [CrossRef]

54. Huang, W.; Zhao, J.; Xing, A.; Wang, G.; Tao, H. Influence of tool path strategies on fatigue performance of high-speed ball-end-milled AISI H13 steel. *Int. J. Adv. Manuf. Technol.* **2018**, *94*, 371–380. [CrossRef]

55. Sedláček, M.; Podgornik, B.; Vižintin, J. Influence of surface preparation on roughness parameters, friction and wear. *Wear* **2009**, *266*, 482–487. [CrossRef]

56. Menezes, P.L.; Kailas, S.V. Effect of surface roughness parameters and surface texture on friction and transfer layer formation in tin–steel tribo-system. *J. Mater. Process. Technol.* **2008**, *208*, 372–382. [CrossRef]

57. Magri, M.L.; Diniz, A.E.; Button, S.T. Influence of surface topography on the wear of hot forging dies. *Int. J. Adv. Manuf. Technol.* **2013**, *65*, 459–471. [CrossRef]

58. Deng, W.J.; Xia, W.; Li, C.; Tang, Y. Formation of ultra-fine grained materials by machining and the characteristics of the deformation fields. *J. Mater. Process. Technol.* **2009**, *209*, 4521–4526. [CrossRef]

59. Umbrello, D. Investigation of surface integrity in dry machining of Inconel 718. *Int. J. Adv. Manuf. Technol.* **2013**, *69*, 2183–2190. [CrossRef]

60. Ramesh, A.; Melkote, S.N. Modeling of white layer formation under thermally dominant conditions in orthogonal machining of hardened AISI 52100 steel. *Int. J. Mach. Tool Manuf.* **2008**, *48*, 402–414. [CrossRef]

61. Elbestawi, M.; Chen, L.; Becze, C.; El-Wardany, T. High-speed milling of dies and molds in their hardened state. *CIRP Ann. Manuf. Technol.* **1997**, *46*, 57–62. [CrossRef]

62. Zhang, S.; Li, W.; Guo, Y. Process design space for optimal surface integrity in finish hard milling of tool steel. *Prod. Eng.* **2012**, *6*, 355–365. [CrossRef]

63. Li, A.; Zhao, J.; Dong, Y.; Wang, D.; Chen, X. Surface integrity of high-speed face milled Ti-6Al-4V alloy with PCD tools. *Mach. Sci. Technol.* **2013**, *17*, 464–482. [CrossRef]

64. Devillez, A.; le Coz, G.; Dominiak, S.; Dudzinski, D. Dry machining of Inconel 718 workpiece surface integrity. *J. Mater. Process. Technol.* **2011**, *211*, 1590–1598. [CrossRef]

65. Pan, Z.; Liang, S.Y.; Garmestani, H.; Shih, D.S. Prediction of machining-induced phase transformation and grain growth of Ti-6Al-4 V alloy. *Int. J. Adv. Manuf. Technol.* **2016**, *87*, 859–866. [CrossRef]

66. Chou, Y.K.; Evans, C.J. White layers and thermal modeling of hard turned surfaces. *Int. J. Mach. Tool Manuf.* **1999**, *39*, 1863–1881. [CrossRef]

67. Swaminathan, S.; Shankar, M.R.; Lee, S.; Hwang, J.; King, A.H.; Kezar, R.F.; Rao, B.C.; Brown, T.L.; Chandrasekar, S.; Compton, W.D.; et al. Large strain deformation and ultra-fine grained materials by machining. *Mater. Sci. Eng. A* **2005**, *410*, 358–363. [CrossRef]

68. Wang, Q.; Liu, Z.; Yang, D.; Mohsan, A. Metallurgical-based prediction of stress-temperature induced rapid heating and cooling phase transformations for high speed machining Ti-6Al-4V alloy. *Mater. Des.* **2017**, *119*, 208–218. [CrossRef]

69. Li, A.; Pang, J.; Zhao, J.; Zang, J.; Wang, F. FEM-simulation of machining induced surface plastic deformation and microstructural texture evolution of Ti-6Al-4V alloy. *Int. J. Mech. Sci.* **2017**, *123*, 214–223. [CrossRef]

70. Thomas, M.; Turner, S.; Jackson, M. Microstructural damage during high-speed milling of titanium alloys. *Scr. Mater.* **2010**, *62*, 250–253. [CrossRef]

71. Shyha, I.; Gariani, S.; El-Sayed, M.A.; Huo, D. Analysis of Microstructure and Chip Formation When Machining Ti-6Al-4V. *Metals* **2018**, *8*, 185. [CrossRef]

72. Li, B.; Zhang, S.; Yan, Z.; Jiang, D. Influence of edge hone radius on cutting forces, surface integrity, and surface oxidation in hard milling of AISI H13 steel. *Int. J. Adv. Manuf. Technol.* **2018**, *95*, 1153–1164. [CrossRef]

73. Liu, B.; Zhou, X.; Hashimoto, T.; Zhang, X.; Wang, J. Machining introduced microstructure modification in aluminium alloys. *J. Alloys Compd.* **2018**, *757*, 233–238. [CrossRef]

74. Aramcharoen, A.; Mativenga, P. White layer formation and hardening effects in hard turning of H13 tool steel with CrTiAlN and CrTiAlN/MoST-coated carbide tools. *Int. J. Adv. Manuf. Technol.* **2008**, *36*, 650–657. [CrossRef]

75. Lu, Y.; Guo, C. Finite element modeling of multi-pass machining of Inconel 718. In Proceedings of the CD 2009 ASME International Conference on Manufacturing Science and Engineering, West Lafayette, IN, USA, 4–7 October 2009; pp. 84081–84086.

76. Wu, G.Q.; Shi, C.L.; Sha, W.; Sha, A.; Jiang, H. Effect of microstructure on the fatigue properties of Ti–6Al–4V titanium alloys. *Mater. Des.* **2013**, *46*, 668–674. [CrossRef]

77. Sealy, M.P.; Guo, Y.B.; Caslaru, R.C.; Sharkins, J.; Feldman, D. Fatigue performance of biodegradable magnesium–calcium alloy processed by laser shock peening for orthopedic implants. *Int. J. Fatigue* **2016**, *82*, 428–436. [CrossRef]

78. Fellah, M.; Samad, M.A.; Labaiz, M.; Assala, O.; Iost, A. Sliding friction and wear performance of the nano-bioceramic α-Al$_2$O$_3$ prepared by high energy milling. *Tribol. Int.* **2015**, *91*, 151–159. [CrossRef]

79. Pang, J. Investigation on Texture Simulation and Friction Characteristics of Machined Surface of Titanium Alloy. Ph.D. Thesis, Shandong University, Shandong, China, 2017.

80. Huang, W.; Zhao, J.; Ai, X.; Wang, G. Influence of tool path strategies on friction and wear behavior of high-speed ball-end-milled hardened AISI D2 steel. *Int. J. Adv. Manuf. Technol.* **2018**, *96*, 2769–2779. [CrossRef]

81. Zhong, Z. Surface Integrity and Corrosion Resistance in High Speed Milling Aluminum Alloy 7050-T7451. Ph.D. Thesis, Shandong University, Shandong, China, 2015.

82. Kim, D.; Kim, B.; Kang, C. Estimation of die service life for a die cooling method in a hot forging process. *Int. J. Adv. Manuf. Technol.* **2005**, *27*, 33–39. [CrossRef]

83. Guo, Y.; Li, W.; Jawahir, I. Surface integrity characterization and prediction in machining of hardened and difficult-to-machine alloys: A state-of-art research review and analysis. *Mach. Sci. Technol.* **2009**, *13*, 437–470. [CrossRef]

84. Arnone, M. *High Performance Machining*; Hanser Gardner Publications: Cincinnati, OH, USA, 1998.

85. Arunachalam, R.; Mannan, M.; Spowage, A. Surface integrity when machining age hardened Inconel 718 with coated carbide cutting tools. *Int. J. Mach. Tool Manuf.* **2004**, *44*, 1481–1491. [CrossRef]

86. Arunachalam, R.; Mannan, M.; Spowage, A. Residual stress and surface roughness when facing age hardened Inconel 718 with CBN and ceramic cutting tools. *Int. J. Mach. Tool Manuf.* **2004**, *44*, 879–887. [CrossRef]

87. Brinksmeier, E.; Cammett, J.; König, W.; Leskovar, P. Residual stresses—Measurement and causes in machining processes. *CIRP Ann. Manuf. Technol.* **1982**, *31*, 491–510. [CrossRef]

88. Sun, J.; Guo, Y. A comprehensive experimental study on surface integrity by end milling Ti–6Al–4V. *J. Mater. Process. Technol.* **2009**, *209*, 4036–4042. [CrossRef]

89. Rao, B.; Dandekar, C.R.; Shin, Y.C. An experimental and numerical study on the face milling of Ti–6Al–4V alloy: Tool performance and surface integrity. *J. Mater. Process. Technol.* **2011**, *211*, 294–304. [CrossRef]

90. Aspinwall, D.K.; Dewes, R.C.; Ng, E.G.; Sage, C.; Soo, S.L. The influence of cutter orientation and workpiece angle on machinability when high-speed milling Inconel 718 under finishing conditions. *Int. J. Mach. Tool Manuf.* **2007**, *47*, 1839–1846. [CrossRef]

91. Jiang, X.; Li, B.; Yang, J.; Zuo, X.Y. Effects of tool diameters on the residual stress and distortion induced by milling of thin-walled part. *Int. J. Adv. Manuf. Technol.* **2013**, *68*, 175–186. [CrossRef]

92. Sridhar, B.; Devananda, G.; Ramachandra, K.; Bhat, R. Effect of machining parameters and heat treatment on the residual stress distribution in titanium alloy IMI-834. *J. Mater. Process. Technol.* **2003**, *139*, 628–634. [CrossRef]

93. Axinte, D.; Dewes, R. Surface integrity of hot work tool steel after high speed milling-experimental data and empirical models. *J. Mater. Process. Technol.* **2002**, *127*, 325–335. [CrossRef]

94. Zhang, S.; Ding, T.; Li, J. Determination of surface and in-depth residual stress distributions induced by hard milling of H13 steel. *Prod. Eng.* **2012**, *6*, 375–383. [CrossRef]

95. Wang, J.; Zhang, D.; Wu, B.; Luo, M. Residual stresses analysis in ball end milling of nickel-based Superalloy Inconel 718. *Mater. Res.* **2017**, *20*, 1681–1689. [CrossRef]

96. Jiang, X.; Li, B.; Yang, J.; Zuo, X.; Li, K. An approach for analyzing and controlling residual stress generation during high-speed circular milling. *Int. J. Adv. Manuf. Technol.* **2013**, *66*, 1439–1448. [CrossRef]

97. Li, B.; Jiang, X.; Yang, J.; Liang, S.Y. Effects of depth of cut on the redistribution of residual stress and distortion during the milling of thin-walled part. *J. Mater. Process. Technol.* **2015**, *216*, 223–233. [CrossRef]

98. Liang, T.; Zhang, D.; Yao, C.; Wu, D.; Zhang, J. Evolution and empirical modeling of compressive residual stress profile after milling, polishing and shot peening for TC17 alloy. *J. Manuf. Process.* **2017**, *26*, 155–165.

99. Fergani, O.; Jiang, X.; Shao, Y.; Welo, T.; Yang, J. Prediction of residual stress regeneration in multi-pass milling. *Int. J. Adv. Manuf. Technol.* **2016**, *83*, 1153–1160. [CrossRef]

100. Wan, M.; Ye, X.Y.; Yang, Y.; Zhang, W. Theoretical prediction of machining-induced residual stresses in three-dimensional oblique milling processes. *Int. J. Mech. Sci.* **2017**, *133*, 426–437. [CrossRef]

101. Yao, C.F.; Tan, L.; Ren, J.X.; Lin, Q. Surface integrity and fatigue behavior for high-speed milling Ti–10V–2Fe–3Al titanium alloy. *J. Fail. Anal. Prev.* **2014**, *14*, 102–112. [CrossRef]

102. Moussaoui, K.; Mousseigne, M.; Senatore, J.; Chieragatti, R. Influence of milling on the fatigue lifetime of a Ti6Al4V titanium alloy. *Metals* **2015**, *5*, 1148–1162. [CrossRef]

103. Yang, D.; Liu, Z. Surface integrity generated with peripheral milling and the effect on low-cycle fatigue performance of aeronautic titanium alloy Ti-6Al-4V. *Aeronaut. J.* **2018**, *122*, 316–332. [CrossRef]

104. McClung, R. A literature survey on the stability and significance of residual stresses during fatigue. *Fatigue Fract. Eng. Mater.* **2007**, *30*, 173–205. [CrossRef]

105. Ghanem, F.; Sidhom, H.; Braham, C.; Fitzpatrick, M. Effect of near-surface residual stress and microstructure modification from machining on the fatigue endurance of a tool steel. *J. Mater. Eng. Perform.* **2002**, *11*, 631–639. [CrossRef]

106. Li, W.; Guo, Y.; Barkey, M.; Jordon, J. Effect tool wear during end milling on the surface integrity and fatigue life of Inconel 718. *Procedia CIRP* **2014**, *14*, 546–551. [CrossRef]

 *applied*
sciences

Article

Fracture Resistance of Monolithic Zirconia Crowns on Four Occlusal Convergent Abutments in Implant Prosthesis

Ting-Hsun Lan [1,2,*], Chin-Yun Pan [3], Pao-Hsin Liu [4] and Mitch M.C. Chou [5]

[1] Division of Prosthodontics, Department of Dentistry, Kaohsiung Medical University Hospital, Kaohsiung Medical University, Kaohsiung 80708, Taiwan
[2] School of Dentistry, College of Dental Medicine, Kaohsiung Medical University, Kaohsiung 80708, Taiwan
[3] Division of Orthodontics, Department of Dentistry, Kaohsiung Medical University Hospital, Kaohsiung Medical University, Kaohsiung 80708, Taiwan
[4] Department of Biomedical Engineering, I-Shou University, Kaohsiung 82442, Taiwan
[5] Department of Materials & Optoelectronic Science, National Sun Yat-Sen University, Kaohsiung 80424, Taiwan
* Correspondence: tinghsun.lan@gmail.com; Tel.: +886-7-3121101 (ext. 2154-11)

Received: 28 April 2019; Accepted: 21 June 2019; Published: 26 June 2019

Featured Application: Thicker and appropriately designed monolithic zirconia crown in the posterior implant area may have better fracture resistance even if the abutment needs to be milled to a convergent occlusal surface for a new insertion pathway.

Abstract: Adjusting implant abutment for crown delivery is a common practice during implant installation. The purpose of this study was to compare the fracture resistance and stress distribution of zirconia specimens on four occlusal surface areas of implant abutment. Four implant abutment designs [occlusal surface area (SA) SA100, SA75, SA50, and SA25] with 15 zirconia prostheses over the molar area per group were prepared for cyclic loading with 5 Hz, 300 N in a servo-hydraulic testing machine until fracture or automatic stoppage after 30,000 counts. The minimum occlusal thickness of all specimens was 0.5 mm. Four finite element models were simulated under vertical or oblique 10-degree loading to analyze the stress distribution and peak value of zirconia specimens. Data were statistically analyzed, and fracture patterns were observed under a scanning electron microscope. Cyclic loading tests revealed that specimen breakage had moderately strong correlation with the abutment occlusal area ($r = 0.475$). Specimen breakage differed significantly among the four groups ($P = 0.001$). The lowest von Mises stress value was measured for prosthesis with a smallest abutment occlusal surface area (SA25) and the thickest zirconia crown. Thicker zirconia specimens (SA25) had higher fracture resistance and lowest stress values under 300 N loading.

Keywords: monolithic zirconia crown; dental implant abutment; cyclic loading; finite element analysis

---

1. Introduction

Implant-supported fixed partial dental prosthesis is a popular treatment option for partially edentulous patients. The mean implant failure rate of fixed partial dentures in partially edentulous patients was reported to be 6% in both the maxilla and mandible (range: 1.9–12.5%) [1,2]. Conditions that contribute to restoration failure include cracking, chipping fracture, bulk fracture, excessive wear of the opposing tooth surface, excessive roughening of the ceramic surface, and unacceptable esthetics. Anusavice [3] defined restoration success as intact prosthesis survival with acceptable surface quality, anatomic contour, function, and esthetic. In clinical settings, delivering an implant prosthesis is challenging when the implant tilts during implantation, which may stem from anatomic limitation, unique treatment design (such as all-on-4), or other iatrogenic factors. Delivery methods of implant crowns mainly divide into the following: Cement-retained and screw-retained implant prosthesis. The implant abutment is always milled or custom-made to fit the insertion pathway.

Fixing a single implant prosthesis on a tilted implant fixture over the posterior area is often a challenging task due to limited insertion pathway, short interdental space, and difficult interproximal hygiene maintenance. After properly milling the implant abutment, one must put into consideration the ceramic materials used, prosthesis thickness, and repairing method after implant prosthesis failed in the future. The advantages of screw-retained implant prosthesis are easy retrieval, better tissue response, and capable of installing when interocclusal space is limited. On the other hand, delivering a cement-retained implant is easier than delivering a screw-retained implant. However, the clinician should evaluate the retention and resistance of the crown meticulously and clean excess cement after delivery.

Based on the method of fabrication, implant abutments can be classified into prefabricated and custom-made abutments. Prefabricated abutments are often used in the esthetic zones owing to difficulty in achieving a good emergence profile, and the margin of the abutment requires milling because it remains supragingival. The milling of prefabricated implant abutment is typically performed using high-speed handpiece and diamond dental burs [4,5]. The total occlusal convergence (TOC) [6], which is the convergence angle between two opposing axial walls, is one of the critical factors for the success of ceramic crowns. Low values of TOC [7,8], range 2° to 5°, are difficult for crown delivery and result in an increasing cement hydraulic pressure. High values of TOC [9], over 24°, are difficult to obtain proper retention of the crown in long-term follow-up. Owing to advances in software and hardware for the open computer numerical control (CNC) system development, computer-aided design (CAD) and computer-aided manufacture (CAM) have increased interoperability and dimension control when milling the implant abutment [10]. Designing abutment using software and milling metal/zirconia material have gained great popularity in clinical dentistry. These abutments have allowed the dentists to create individual emergence profile and provide an ideal structure and crown path insertion. After manual or machine milling, the morphology of the implant abutment can be adjusted, and the prosthesis is always set on an asymmetrical convergent angle of the axial surfaces which lead to different occlusal areas.

Zirconia is a crystalline form of zirconium dioxide; it has excellent mechanical properties similar to metal and with color close to natural teeth. The best properties for medical use are exhibited by zirconia stabilizing with Y_2O_3—also known as tetragonal zirconia polycrystal (TZP) [11]. A high flexure strength (900–1200 MPa) and compression resistance (2000 MPa) [12] render it suitable for use in clinical dentistry [13]. Monolithic zirconia crowns have gained popularity since 2010 [14–16], primarily for reducing the major clinical complications that resulted from the fracture of porcelain veneers and zirconia frameworks [17]. Lawn et al. [18] mentioned that failure mode, materials, flaw state, layer thickness, and cement may affect the fracture resistance of ceramic crowns. Deng et al. [19] showed that fully dense zirconia (Y-TZP) has the highest critical load (100–1000 N) at low thicknesses (0.1–1.0 mm) when comparing different ceramic specimens. Hamburger et al. [20] showed that the fracture risk of ceramic materials has a positive correlation to layer thickness. Weigl et al. [21] showed that 0.5 mm thick monolithic crowns possessed sufficient strength to endure physiological performance, regardless of the type of cementation. Lan et al. [22] showed that the mean fracture loading cycles of a 0.5-mm zirconia specimen is 8480 ± 2009. He recommended the use of an implant prosthesis of monolithic zirconia crown with a thickness of 0.8 mm during clinical practice. Additionally, Kelly [23] suggested a more relevant approach using the finite element method (FEM) to solve for stress as a function of load. Anatomically, axial loading posterior teeth with the implant prosthesis can vary from 42 to 412 N [24], and Raadsheer et al. [25] found that maximum voluntary occlusal force was 545.7 N in men and 383.6 N in women in natural dentition. On the other hand, Lan et al. [22] used 300 N to simulate the bite force of implant prosthesis of molar area. During maximum intercuspation

(MICP), loading force would spread among all contacted teeth. Therefore, we anticipated that maximal voluntary occlusal force should not approach the value Raadsheer has presented.

As discussed above, the fracture resistance of monolithic zirconia specimens is primarily affected by the thickness [22], and researchers have attempted to determine the correlation between the different convergent occlusal area of implant abutment and breaking test of zirconia specimens after lead bearing. Therefore, the purpose of this study is to compare the fracture resistance and stress distribution of zirconia specimens on four different occlusal convergence implant abutments. The hypothesis is that more occlusal convergence of implant abutment would lead to less fracture resistance of monolithic zirconia crowns.

2. Materials and Methods

2.1. Specimens Preparation

Figure 1 shows four implant abutments [occlusal surface area (SA) SA100, SA75, SA50, and SA25] in the molar area that were prepared to receive monolithic zirconia specimens. SA50 implies that the total occlusal surface contact area is half of SA100. The removed occlusal surface area (A) was calculated using the equation:

$$A = \pi r^2 \cdot \theta/360 - 1/2\ r^2 \sin\theta \tag{1}$$

where r is the radius of the circle; θ is the central angle measured in radians.

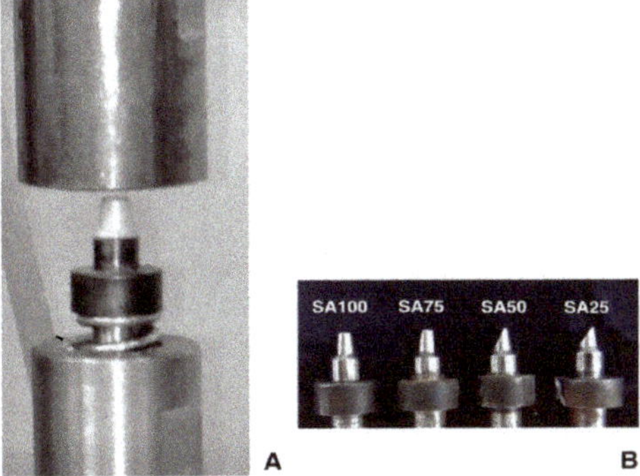

Figure 1. (**A**) Zirconia specimen and holder with spring to simulate human tissue by the servo-hydraulic testing machine; (**B**) four implant-abutment convergent occlusal surface areas. From left to right, SA100, SA75, SA50, and SA25. SA stands for the occlusal surface area, SA100 is the standard implant abutment, SA75 means 75% of occlusal surface area relative to SA100, and so on.

The specimen holder and the operating program of the test machine were adjusted to simulate the human masticatory cycle. The specimen holder had three parts: A precise adapter for the test machine, a heat-treated steel rod with a tapered tip to accommodate the specimen, and a spring to simulate cyclic occlusal contact and position control. The spring was checked before and after tests to verify the slope of strain and stress by Hooke's law [26], as it should remain constant throughout tests. The G power analysis [27] was used to estimate the required sample size; assuming four test groups, an effect size of 0.56, the probability of an α error of 0.05, and power of 0.95. The sample size was set to have 15 per group. Overall, 60 complete zirconia specimens were designed using the computer-aided design/computer-aided manufacturing (CAD/CAM) technique to match the implant abutment design. The specimens were designed by the DentalCAD software (Exocad 2016, Exocad GmbH, Darmstadt, Germany). All specimens were milled from one commercial block of Y-TZP zirconia V (made in made in Bad Sackingen, Germany) using the same open CNC system milling machine (ARDENTA CNC MILL, CS100-5A, ARIX, Tainan, Taiwan) and were densely sintered at 1450 °C for 2 h.

2.2. Cyclic Loading Test

Figure 2 is the schematic representation of four implant abutment designs. The specimens were retained on the implant dies by friction and could only be removed by crown remover. The stability of the prosthesis was enhanced by the anti-rotation surface design of the dental implant abutment. The dimension of SA100 specimen was designed to fit the implant abutment with thickness of 0.5 mm over occlusal surface, diameter of 5 mm at base, and diameter of 3.5 mm at top. The reduction began 2 mm from the bottom to preserve retentive property, and tapered up to the desired occlusal surface area. Reduced volume was replaced with zirconia. Fifteen specimens in each group were tested vertically at 5 Hz and 300 N in a servo-hydraulic testing machine (Instron M8810, Instron Ltd. Norwood, MA, USA) until the specimen fractured or the test machine automatically stopped after 30,000 cyclic counts. To ensure that a consistent force was exerted on the specimen, the computer program had a 2-stage control to emulate the masticatory cycle of human jaw. The first stage, which simulated natural occlusal displacement, was position control. This stage ensured constant stroke distance between the load cell and specimen. The second stage was load control. This stage ensured a consistent force exerted by the load cell to the specimen remained constant throughout the test cycle.

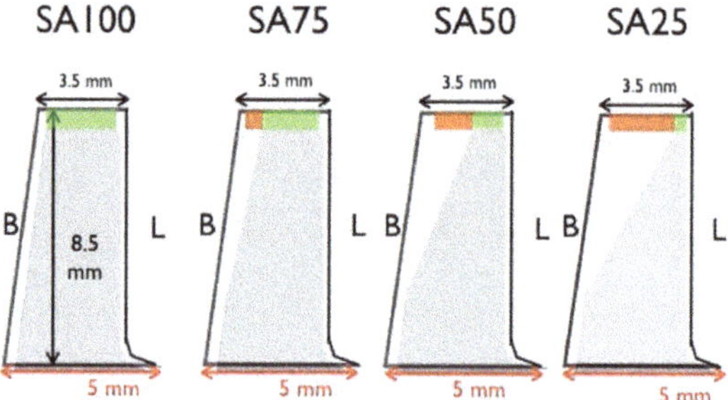

Figure 2. Schematic representation of the dimension of zirconia specimens on four implant abutments. B means buccal, L means lingual; Green color represents the occlusal surface area connecting the zirconia specimen and implant abutment. Brown color represents the implant abutment occlusal surface area reduction replaced with zirconia material.

The indentation stress-strain relation is well defined by the classic Hertzian theory for ideally elastic, homogeneous bulk materials with Young's modulus E and Poisson's ratio ν. The linear form of the Hertzian solution [28,29] is expressed as:

$$a = \sqrt{rd} \tag{2}$$

$$p = 3F/2\pi a^2 \tag{3}$$

where a is the radius of the contact area, r is the radius of indent and d the depth, F is the loading, and p is the maximum contact pressure.

2.3. Finite Element Method

The dental implant abutment models were constructed using the CAD software (SolidWorks 2010, SolidWizard Corporation, Concord., MA, USA). Four finite element (FE) models with the different occlusal surface areas covered by zirconia specimens were constructed to simulate the posterior molar region. ANSYS was the FEM software used in this study (ANSYS Workbench 11, Canonsburg, PA, USA). The Young's modulus and Poisson ratio of Ti-6Al-4V [30] were 110 GPa and 0.33, respectively. The Young's modulus and Poisson ratio of zirconia [31] were 220 GPa and 0.3, respectively.

The materials used in the models were assumed to be isotropic, homogeneous, and linearly elastic. The average tetrahedral element size was 1 mm in length and the 3D FE model of SA100 model was meshed in the 115484 elements and 147842 nodes by convergence test. The 10-node tetrahedral element could provide six degrees of freedom (DOF), including three translations and three rotations on a node. Hence, total number of the DOF in the FEM was related to real nodal numbers. The boundary constrain of the bottom of the holder base would restrict the DOF of the implant-prosthesis complex of any rotational or translational movement. The mesh convergence test of element sensitivity was performed in this study before FEM. The convergence criterion was set to 5% to identify a reliable numerical result of the FEM.

The contact analysis in the contact area must be assumed that contact and target elements were paired to analyze surface interaction between two solid models under loading. Furthermore, the coefficient of friction must also be applied in the contact area. However, in this study, our main goal was to test the fracture resistance of different zirconia crown and implant abutment designs. Therefore, the sliding effect between the zirconia crown and implant abutment was considered negligible. For contact connection between the zirconia crown and implant abutment, we defined the finite element analysis (FEA) as a non-separating connection without any frictional effect. Hence, the contact analysis with frictional effect was not included in this study. Vertical loading of 300 N was applied on the occlusal surface of the specimens, and a 4-directional oblique force of 300 N at 10 degrees was applied to the marginal ridge of the specimens. The maximum von Mises (EQV) stress values were measured for specimens in different loading directions and implant prosthesis and abutment designs.

2.4. Statistical Analysis and Microstructural Observation

Data were normally distributed (P > 0.05) and there was no homogeneity of variance among groups of tests (P > 0.05). Data were compared using the Krukal-Wallis test, and the Spearman correlation was determined using a statistical program (SPSS Statistics for Windows, v20; IBM Corp, Armonk, NY, USA). Using scanning electron microscope imaging (JSM-6360; JEOL, Tokyo, Japan), the microscopic conditions of the specimens were examined.

3. Results

3.1. Cyclic Load Test

Table 1 showed the mean cyclic number and broken number/total number of specimens tested until fracture. The result showed 11061.6 (11/15), 12278 (6/15), 15617 (5/15), and 8000 (1/15) for the implant abutments with occlusal surface areas of SA100, SA75, SA50, and SA25, respectively.

Table 1. Results of zirconia specimens on four occlusal surface areas of implant abutment after cycling test under vertical 300 N loading.

Abutment Occlusal Surface Area (%)	Total Cycles Mean ± SD	Broken Specimens/ Total Specimens
SA100	11,061.6 ± 4602.6 [a]	11/15
SA75	12,278 ± 4139.3 [abc]	6/15
SA50	15,617 ± 3508.4 [bc]	5/15
SA25	8000 [bc]	1/15
P value *	0.001	<0.01

Different superscript letters in the column represented statistical significance between groups (P < 0.05; post hoc Dunn test). Total cycles indicated total number of cycles ran before specimens broke under 300 N loading or 30,000 cycles if specimens remained intact. SD: Standard deviation. a, b, c are notations depicting different class variations.

*Krukal-Wallis test (K independent samples).

Spearman's correlation coefficients ($r = 0.475$) in association with different abutment designs and occlusal surface areas to fracture resistance of zirconia crown.

The broken specimens of four occlusal surface areas showed specimens breaking into two to three fragments (Figure 3A–D). The cyclic number of broken specimens showed a significant difference among different groups of specimens (P = 0.001). Dunn's honestly significant difference test indicated that the specimen breakage of SA100 was statistically significant when comparing to SA50 and SA25 (P = 0.001). Figure 4 showed that the percentage of intact specimens were 27% (SA100), 60% (SA75), 67% (SA50), and 93% (SA25), without visible fracture line found after 30,000 cycles. Specimen breakage had a moderately strong association with the implant abutment occlusal surface area (r = 0.475).

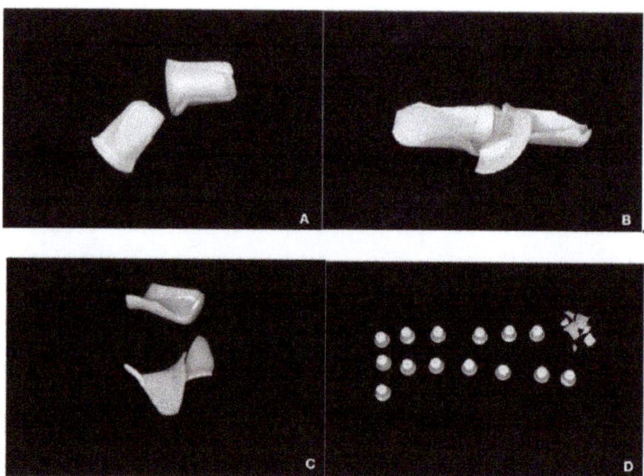

Figure 3. Specimens on four implant abutment designs after testing. A, the SA100 specimen was broken around the top and axial wall corner with apparently broken line. B, the SA75 specimen was broken into three segments. C, the SA50 specimen with fractured segments. D, one zirconia specimen on the SA25 abutment design was broken into multiple segments, and the remaining specimens's morphology remained intact.

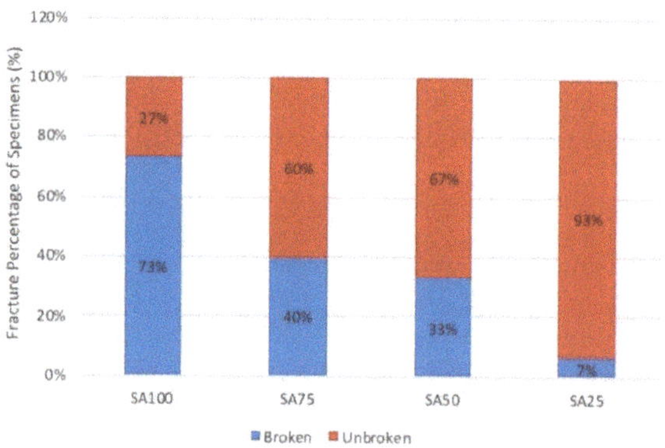

Figure 4. Histogram of results shows the percentage of broken and unbroken specimens corresponding to four types of abutment designs SA100, SA75, SA50, SA25.

3.2. Finite Element Analysis

Figure 5 was a cross-section view of the SA25 model experiencing vertical loading of 300 N. The stress was transmitted from the specimen to the implant abutment and the maximum stress value was shown on the lingual edge of the specimens. Figures 6 and 7 showed that stress was concentrated around the corner area on the occlusal axial wall of the specimens where forces were applied vertically or obliquely. The maximum EQV was higher under oblique loading compared to vertical loading. Figure 8 compared the maximum EQV value of four implant abutment designs under vertical and oblique 10-degree 300 N loading. The results showed that the peak EQV of SA25 had the lowest value compared to the other three types. Oblique lingual loading had the lowest peak EQV out of all oblique loading directions. The percentage of fracture specimens among the four groups was positively associated with the results of FEA ($r = 0.391$).

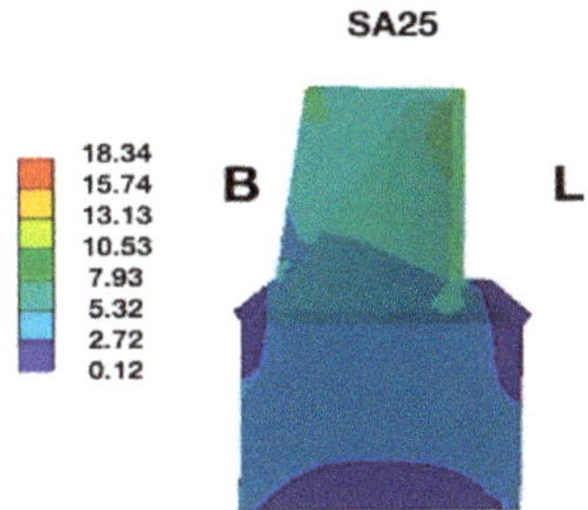

Figure 5. A cross-section of the finite element (FE) model with specimen set on the implant abutment. Using SA25 as an example, stress distribution of the whole model was shown. The maximum stress value was located on the occluso-lingual axial wall of the specimens when vertical forces were applied.

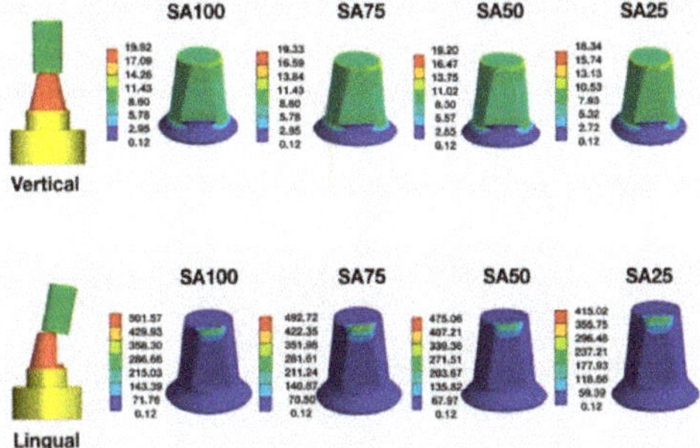

Figure 6. Distribution of stress in zirconia specimens. (Top) The vertical load of 300 N from lingual (L) view. (Bottom) Oblique 10-degree load of 300 N on the lingual marginal ridge. From the left to right are loading models: SA100, SA75, SA50, and SA25.

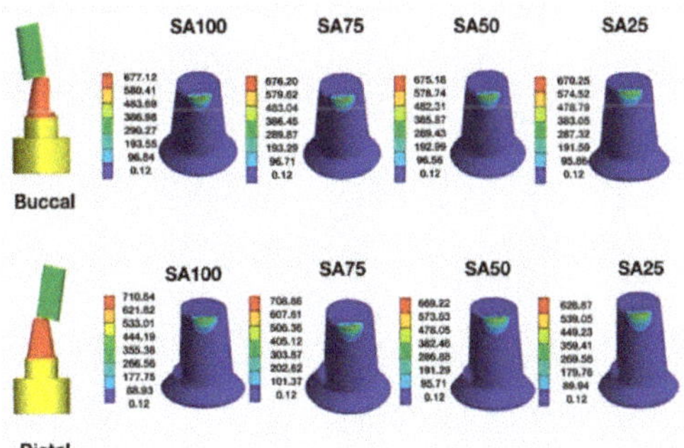

Figure 7. Distribution of stress in zirconia specimens. (Top) Oblique 10-degree load of 300 N on the buccal marginal ridge. (Bottom) Oblique 10-degree load of 300 N on the buccal marginal ridge. From the left to right are loading models: SA100, SA75, SA50, and SA25.

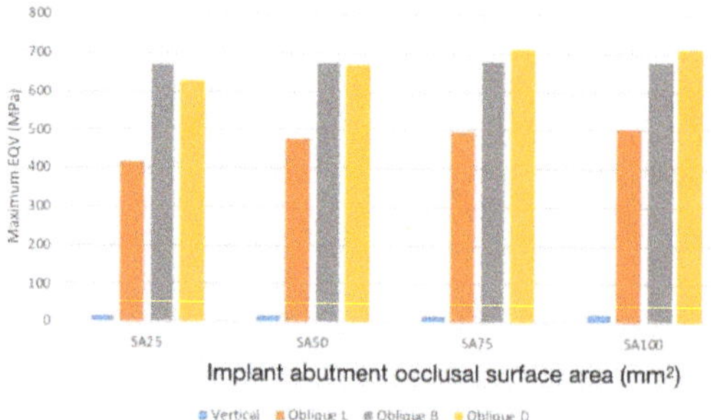

Figure 8. Comparison of maximum von Mises (EQVs) in zirconia specimens on four implant abutment designs (SA25, SA50, SA75, and SA100) under vertical (300 N), oblique L, oblique B, and oblique D directions. Lowest value occurred in the zirconia specimen with the implant occlusal surface of SA25. Oblique L means oblique 10-degree 300 N loading applied on a lingual marginal ridge of zirconia specimen. L means lingual. B means buccal. D means distal.

3.3. Microstructure of Zirconia Block After Cyclic Loading Test

Figure 9 showed the microstructure of the fractured zirconia segment after loading when viewed with scanning electron microscope (SEM). The SEM images in Figure 9A showed beach marks, which indicated the progressive fatigue failure of the zirconia specimen. Figure 9B,C revealed an apparent fracture line visible alongside the beach marks when a brittle failure of the material occurred. Figure 9D showed full densification and sufficient sintering of the zirconia and presence of some small pores at grain and grain boundaries.

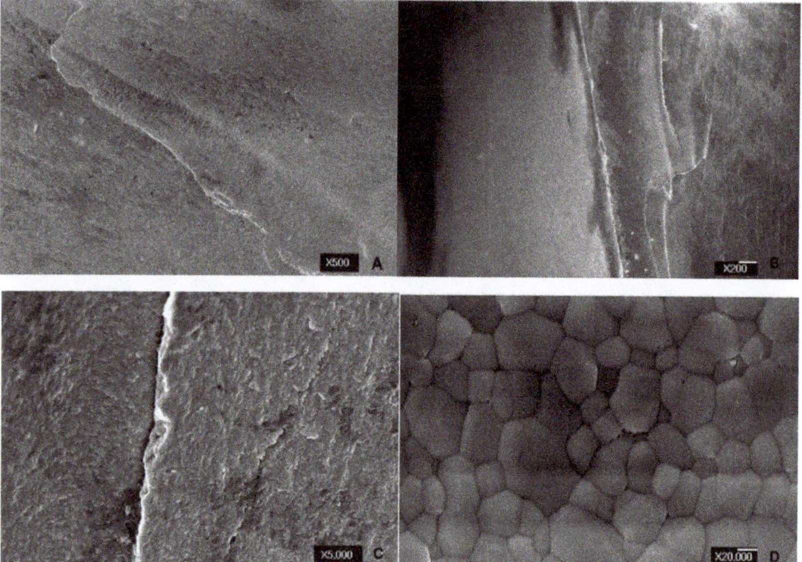

Figure 9. Scanning electron microscopic images showing fracture condition of broken segments. (**A,B**), beach marks show fracture segments from the specimen on SA25 (original magnification ×500) and SA50 (original magnification ×200). (**C**), apparent cracks observed on SA50 fracture segments (original magnification ×5000). (**D**), intact surface of zirconia specimens (original magnification ×20,000).

4. Discussion

Lan et al. [22] found that the fracture resistance of zirconia specimens have positive correlation with prosthesis thickness, and the mean cyclic number of broken specimens on 0.5 mm thickness was 8480 ± 2009. This study demonstrated breakage in 11 of 15 specimens (73%) and a similar cyclic number of broken specimens for 0.5 mm (SA100) zirconia specimens (Table 1). In addition, the breakage percentage of SA75 specimens (6/15) showed no statistical significance (P > 0.05) compared to SA50 specimens (5/15). On the other hand, only one SA25 zirconia specimen (1/15) broke during 300 N vertical loading. The results supported the speculation that the implant abutment occlusal surface area affects the fracture resistance of the zirconia crown, although only a moderately strong correlation was found ($r = 0.475$). The FEM showed the lowest peak stress value for SA25 specimen in all loading types, hence, we speculated that the implant abutment occlusal area has an inverse proportion to fracture resistance of the zirconia specimens (Figures 1 and 2). Under oblique loading, the peak value of the crown from the lingual side was lower than that from the buccal and distal side. This difference was attributed to the contact area of the prosthesis because the radius of the contact area from the lingual side was greater than other surfaces (Figure 5 bottom row). Moreover, the radius of the contact area negatively affecting the stress under constant loading is clarified by the Hertzian theory [28,29]. On the other hand, the oblique buccal loading and oblique distal loading showed similar peak stress, which may be attributed to the same contact area between the indenter and zirconia crown (Figure 6 top and bottom row).

Cyclic loadings could cause an accumulation of deformation and decrease the load resistance of the tested materials. Static loadings were loads with a very low rate of change to maximum stresses on specimens. Studart et al. [32] showed the cyclic fatigue behavior of zirconia at loads between 50 and 200 N and indicated that zirconia can tolerate the severe cyclic loading in the molar area. On the other hand, static FEM analysis was shown the maximum stress value and stress distribution from an applied load 250 N on the specimens. For brittle materials, Choi and Salem [33] mentioned the prediction of cyclic fatigue lifetime from static lifetime by using the ratio of static to cyclic fatigue life. These results demonstrated the similar failure trend between static and cyclic fatigue for brittle materials, and the fracture of brittle materials would be induced immediately when the maximum stress was higher than the flexural strength. In addition, in this present study, the cyclic experimental tests showed the positive correlation between the percentage of broken specimens over total specimens with the peak stress value of FEA corresponding to four types of abutment designs ($r = 0.391$). Moreover, both cyclic experimental tests and static FEM analysis were shown the same results that the SA25 specimens had the lowest stress values and higher fracture resistance under 300 N loading. The results hence rejected the hypothesis described in the introduction. Alammari et al. [34] reported that an increase of TOC (from 12 to 20 degree) did not affect marginal and internal fit but resulted in higher load to fracture values of ceramic crowns. A similar finding was reported by Corazza et al. [35], who compared the influence of 6°, 12°, and 20° TOC on the fracture load of yttria partially stabilized zirconia veneered with feldspathic porcelain. The 20° TOC had the best outcome. On the other hand, it could be reasonably speculated that the SA25 specimen allowed for greater implant body tilting. Asymmetric abutment milling with different occlusal convergence was inevitable in clinical practice and also needs to put in consideration of proper crown morphology to avoid food deposition, which relied heavily on accurate standard abutment milling and CAD/CAM custom abutment design.

Milling typically increased the sharp angle and surface roughness. The sharp angle without a fitting contact surface with ceramic could easily increase the stress concentration between zirconia and the implant abutment. Therefore, using zirconia to match the abutment occlusal surface can increase the fracture resistance and ensure a proper fitting contact surface; however, if the surface is not finished or polished properly to decrease the improper contact surface, stress would concentrate on the sharp edges between the zirconia and abutment. The breakage of one SA25 zirconia specimen after 8000 cycles under 300 N loading was attributed to the tiny sharp contact point between zirconia and the abutment despite a meticulous inspection before testing.

Implant abutment milling or adjustment was a frequent practice in clinical dentistry. The success of implant restoration and its long-term survival depended on several factors. First, a well-designed treatment plan would put implant installation and restoration into consideration before the operation. Second, the dentists need to confirm whether the implant requires tilting using the abutment design based on the biomechanical and biomaterial aspects. Third, the zirconia crown used should be fully densified and sufficiently sintered for strength and longevity. Fourth, accurate interdental space calculation, meticulous execution of the procedure, and routine follow-up are required to achieve excellent results.

5. Conclusions

The fracture resistance of four implant abutment occlusal convergent areas (SA100, SA75, SA50, and SA25) of the zirconia sintered at low temperatures for dental use was studied using a servo-hydraulic testing machine, FE models, and SEM. Fracture resistance was estimated based on whether or not the specimen broke, and the results were as follows: Breakage was observed in 11 of the 15 specimens with the occlusal surface area of SA100, six of 15 specimens with the occlusal surface area of SA75, five of 15 specimens with the occlusal surface area of SA50, and one of 15 specimens with the occlusal surface area of SA25. The FE model showed that SA25 specimens had the lowest peak EQV value in different loading directions. Thicker zirconia specimens (SA25) had higher fracture resistance and lowest stress values under 300 N loading.

Author Contributions: Conceptualization, T.-H.L. and C.-Y.P.; Data curation, C.-Y.P. and P.-H.L.; Formal analysis, T.-H.L. and P.-H.L.; Investigation, T.-H.L., C.-Y.P., P.-H.L. and M.M.C.C.; Software, C.-Y.P. and P.-H.L.; Supervision, T.-H.L.; Writing—original draft, T.-H.L.; Writing—review and editing, T.-H.L., C.-Y.P., P.-H.L. and M.M.C.C.

Funding: The authors gratefully acknowledge the support of NSYSU-KMU JOINT RESEARCH PROJECT (NSYSUKMUP014, NSYSUKMU103-P003) and partial support of Southern Taiwan Science Park Bureau, Ministry of Science and Technology (CZ-01-01-05-105).

Conflicts of Interest: The authors declare no conflict of interest.

References

1. Misch, C.E.; Goodacre, C.J.; Finley, J.M.; Misch, C.M.; Marinbach, M.; Dabrowsky, T.; English, C.E.; Kois, J.C.; Cronin, R.J., Jr. Consensus conference panel report: Crown-height space guidelines for implant dentistry—Part 1. *Implant Dent.* **2005**, *14*, 312–321. [CrossRef] [PubMed]
2. Misch, C.E.; Goodacre, C.J.; Finley, J.M.; Misch, C.M.; Marinbach, M.; Dabrowsky, T.; English, C.E.; Kois, J.C.; Cronin, R.J., Jr. Consensus conference panel report: Crown-height space guidelines for implant dentistry—Part 2. *Implant Dent.* **2006**, *15*, 113–121. [CrossRef] [PubMed]
3. Anusavice, K.J. Standardizing failure, success, and survival decisions in clinical studies of ceramic and metal–ceramic fixed dental prostheses. *Dent. Mater.* **2012**, *28*, 102–111. [CrossRef] [PubMed]
4. Siegel, S.C.; Fraunhofer, J.A. Dental cutting with diamond burs: Heavy-handed or light-touch? *J. Prosthodont.* **1999**, *8*, 3–9. [CrossRef] [PubMed]
5. Siegel, S.C.; von Fraunhofer, J.A. Cutting efficiency of three diamond bur grit sizes. *J. Am. Dent. Assoc.* **2000**, *131*, 1706–1710. [CrossRef]
6. Prosthodontics, T. The glossary of prosthodontic terms ninth edition (GPT-9). *J. Prosthet. Dent.* **2017**, *117*, e25.
7. Beuer, F.; Edehoff, D.; Gernet, W.; Naumann, M. Effect of preparation angles on the precision of zirconia crown copings fabricated by CAD/CAM system. *Dent. Mater. J.* **2008**, *27*, 814–820. [CrossRef]
8. Hmaidouch, R.; Neumann, P.; Mueller, W. Influence of preparation form, luting space setting and cement type on the marginal and internal fit of CAD/CAM crown copings. *Int. J. Comput. Dent.* **2011**, *14*, 219–226.
9. Shillingburg, H.T.; Sather, D.A.; Wilson, E.L.; Cain, J.; Mitchell, D.; Blanco, L.; Kessler, J. *Fundamentals of Fixed Prosthodontics*, 4th ed.; Quintessence Publishing Company: Chicago, IL, USA, 2007; pp. 299–308.
10. Zhou, L.; Yuan, J.; Gao, P.; Ren, Y. A new architecture of open CNC system based on compiling mode. *Int. J. Adv. Manuf. Technol.* **2014**, *73*, 1597–1603. [CrossRef]
11. Gupta, T.; Bechtold, J.; Kuznicki, R.; Cadoff, L.; Rossing, B. Stabilization of tetragonal phase in polycrystalline zirconia. *J. Mater. Sci.* **1977**, *12*, 2421–2426. [CrossRef]

12. Piconi, C.; Maccauro, G. Zirconia as a ceramic biomaterial. *Biomaterials* **1999**, *20*, 1–25. [CrossRef]

13. Manicone, P.F.; Iommetti, P.R.; Raffaelli, L. An overview of zirconia ceramics: Basic properties and clinical applications. *J. Dent.* **2007**, *35*, 819–826. [CrossRef] [PubMed]

14. Rojas-Vizcaya, F. Full zirconia fixed detachable implant-retained restorations manufactured from monolithic zirconia: Clinical report after two years in service. *J. Prosthodont.* **2011**, *20*, 570–576. [CrossRef] [PubMed]

15. Beuer, F.; Stimmelmayr, M.; Gueth, J.-F.; Edelhoff, D.; Naumann, M. In vitro performance of full-contour zirconia single crowns. *Dent. Mater.* **2012**, *28*, 449–456. [CrossRef]

16. Park, J.-H.; Park, S.; Lee, K.; Yun, K.-D.; Lim, H.-P. Antagonist wear of three CAD/CAM anatomic contour zirconia ceramics. *J. Prosth. Dent.* **2014**, *111*, 20–29. [CrossRef] [PubMed]

17. Schmitter, M.; Mueller, D.; Rues, S. Chipping behaviour of all-ceramic crowns with zirconia framework and CAD/CAM manufactured veneer. *J. Dent.* **2012**, *40*, 154–162. [CrossRef]

18. Lawn, B.R.; Deng, Y.; Thompson, V.P. Use of contact testing in the characterization and design of all-ceramic crownlike layer structures: A review. *J. Prosth. Dent.* **2001**, *86*, 495–510. [CrossRef]

19. Deng, Y.; Lawn, B.R.; Lloyd, I.K. Characterization of damage modes in dental ceramic bilayer structures. *J. Biomed. Mater. Res.* **2002**, *63*, 137–145. [CrossRef]

20. Hamburger, J.; Opdam, N.; Bronkhorst, E.; Huysmans, M. Indirect restorations for severe tooth wear: Fracture risk and layer thickness. *J. Dent.* **2014**, *42*, 413–418. [CrossRef]

21. Weigl, P.; Sander, A.; Wu, Y.; Felber, R.; Lauer, H.-C.; Rosentritt, M. In-vitro performance and fracture strength of thin monolithic zirconia crowns. *J. Adv. Prosthodont.* **2018**, *10*, 79–84. [CrossRef]

22. Lan, T.-H.; Liu, P.-H.; Chou, M.M.; Lee, H.-E. Fracture resistance of monolithic zirconia crowns with different occlusal thicknesses in implant prostheses. *J. Prosthet. Dent.* **2016**, *115*, 76–83. [PubMed]

23. Kelly, J.R. Clinically relevant approach to failure testing of all-ceramic restorations. *J. Prosthet. Dent.* **1999**, *81*, 652–661. [CrossRef]

24. Stanford, C.M.; Brand, R.A. Toward an understanding of implant occlusion and strain adaptive bone modeling and remodeling. *J. Prosthet. Dent.* **1999**, *81*, 553–561. [CrossRef]

25. Raadsheer, M.; Van Eijden, T.; Van Ginkel, F.; Prahl-Andersen, B. Contribution of jaw muscle size and craniofacial morphology to human bite force magnitude. *J. Dent. Res.* **1999**, *78*, 31–42. [CrossRef] [PubMed]

26. Serna, J.D.; Joshi, A. Studying springs in series using a single spring. *Phys. Educ.* **2011**, *46*, 33. [CrossRef]

27. Prajapati, B.; Dunne, M.; Armstrong, R. Sample size estimation and statistical power analyses. *Optom. Today* **2010**, *16*, 10–18.

28. Timoshenko, S.; Goodier, J. *Theory of Elasticity*; McGraw-Hill: New York, NY, USA, 1951; pp. 372–377.

29. Lawn, B.R. Indentation of ceramics with spheres: A century after Hertz. *J. Am. Ceram. Soc.* **1998**, *81*, 1977–1994. [CrossRef]

30. Collings, E. *The Physical Metallurgy of Titanium Alloys*; ASM American Society for Metals: Metals Park, OH, USA, 1984; pp. 201–209.

31. Green, D.J.; Hannink, R.H.; Swain, M.V. *Transformation Toughening of Ceramics*; CRC Press: Boca Raton, FL, USA, 1989; pp. 126–138.

32. Studart, A.R.; Filser, F.; Kocher, P.; Gauckler, L.J. Fatigue of zirconia under cyclic loading in water and its implications for the design of dental bridges. *Dent. Mater.* **2007**, *23*, 106–114. [CrossRef]

33. Choi, S.R.; Salem, J.A. Cyclic fatigue of brittle materials with an indentation-induced flaw system. *Mater. Sci. Eng. A* **1996**, *208*, 126–130. [CrossRef]

34. Alammari, M.R.; Abdelnabi, M.H.; Swelem, A.A. Effect of total occlusal convergence on fit and fracture resistance of zirconia-reinforced lithium silicate crowns. *Clin. Cosmet. Investig. Dent.* **2019**, *11*, 1–8. [CrossRef]

35. Corazza, P.H.; Feitosa, S.A.; Borges, A.L.S.; Della Bona, A. Influence of convergence angle of tooth preparation on the fracture resistance of Y-TZP-based all-ceramic restorations. *Dent. Mater.* **2013**, *29*, 339–347. [CrossRef] [PubMed]

 *applied sciences*

Article

Fatigue Crack Growth under Non-Proportional Mixed Mode Loading in Rail and Wheel Steel Part 1: Sequential Mode I and Mode II Loading

Makoto Akama

Department of Mechanical Engineering for Transportation, Osaka Sangyo University, Osaka 574-8530, Japan; akama@tm.osaka-sandai.ac.jp; Tel.: +81-72-875-3001

Received: 29 March 2019; Accepted: 13 May 2019; Published: 16 May 2019

Abstract: Fatigue tests were performed to estimate the coplanar and branch crack growth rates on rail and wheel steel under non-proportional mixed mode I/II loading cycles simulating the load on rolling contact fatigue cracks; sequential and overlapping mode I and II loadings were applied to single cracks in the specimens. Long coplanar cracks were produced under certain loading conditions. The fracture surfaces observed by scanning electron microscopy and the finite element analysis results suggested that the growth was driven mainly by in-plane shear mode (i.e., mode II) loading. Crack branching likely occurred when the degree of overlap between these mode cycles increased, indicating that such degree enhancement leads to a relative increase of the maximum tangential stress range, based on an elasto–plastic stress field along the branch direction, compared to the maximum shear stress. Moreover, the crack growth rate decreased when the material strength increased because this made the crack tip displacements smaller. The branch crack growth rates could not be represented by a single crack growth law since the plastic zone size ahead of the crack tip increased with the shear part of the loading due to the T-stress, resulting in higher growth rates.

Keywords: non-proportional mixed mode loading; fractography; mode II stress intensity factor; finite element analysis; rail steel; wheel steel

---

1. Introduction

When railway wheels repeatedly pass over the rails, rolling contact fatigue (RCF) cracks occur on both the wheel tread and the railhead. The accumulation of large plastic shear strains generates these cracks, which initially grow in the plastic flow direction of the resulting microstructure [1]; when their growth reaches a depth of several millimeters, they are unaffected by the plastic deformation layer and are then subjected to a mixed mode I/II/III loading, growing as coplanar cracks at shallow angles to the surface [2–5]. The mode II/III loading cycle is almost proportional, while the mode I/II and I/III cycles are non-proportional.

Fremy et al. [6] conducted experiments in non-proportional mixed mode I/II/III loading conditions to investigate the load path effect on fatigue crack growth (FCG) in 316L stainless steel. They applied the 'star' load path to a cruciform specimen containing an inclined crack by using a six-actuator servo-hydraulic testing machine; the addition of mode III loading steps to a mode I/II loading sequence increased the FCG rate even when the crack path was not significantly modified. However, the tested material was not rail or wheel steel and the load history did not simulate the one experiencing RCF cracks. Moreover, only a coplanar crack growth below 2 mm was obtained.

Many studies focused on the coplanar crack growth of rail steel under non-proportional mixed mode I/II loading cycles. Figure 1 shows how mode I and II loadings generated in the rail and wheel when the fluid penetrates into the crack. Details will be described in the next section. Bold et al. [7] performed fatigue tests, sequentially applying cyclic mode I and II loadings on rail steel biaxial cruciform specimens. When they removed the mode I loading before applying the fully reversed mode II and the ratio between the nominal range values of the mode I and II stress intensity factors (SIFs), K_I (ΔK_I), and K_{II} (ΔK_{II}), were greater than or equal to 0.5, a long coplanar crack growth was observed. Wong et al. [8,9] carried out sequential mixed mode I/II loading experiments on cruciform specimens; when they applied mode II loading before removing mode I, a degree of overlap (δ) appeared. The effective values of ΔK_I and ΔK_{II} (ΔK_{Ieff} and ΔK_{IIeff}, respectively) and the δ effect on the crack growth were considered and crack growth models were proposed in the ΔK_{Ieff} and ΔK_{IIeff} forms. They reported that ΔK_{Ieff} is a control parameter determining the crack growth direction and that the δ increase encourages crack branching; they also proposed a branch criterion in terms of ΔK_{Ieff} and δ. Akama and Susuki [10] conducted non-proportional mixed mode I/II FCG tests on rail and wheel steels in practical use in Japan. Based on the study of Wong et al., they defined crack growth models using ΔK_{Ieff} and ΔK_{IIeff} and found that the cracks branched easily in wheel steel compared to rail steel. They also examined the fracture surfaces, but no clear striation pattern was observed.

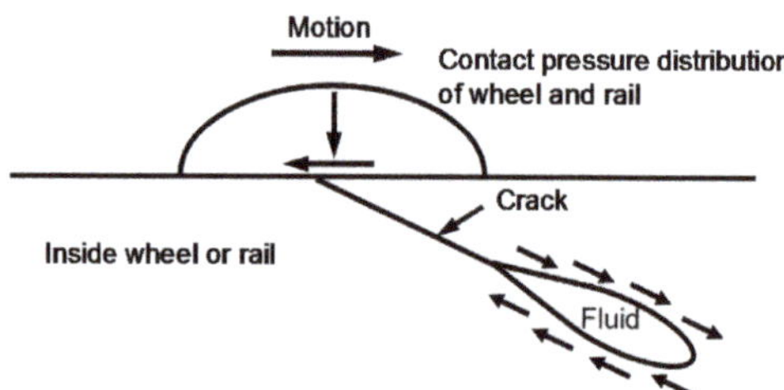

Figure 1. Schematic representation of how a fluid in the crack generates mode I and II loadings.

Sequential mode I/II FCG tests on rail steel were also performed by Doquet and Pommier [11]. They used tubular specimens and gold microgrids with a 4 mm pitch, which were laid ahead of the crack tips and along the crack surfaces to observe the residual shifts; hundreds of microns of constant-speed coplanar growth were obtained when $\Delta K_{II}/\Delta K_I$ ranged from 1 to 4. They also investigated the role of plasticity and crack face friction via finite element analysis (FEA) and concluded that the maximum growth rate criterion rationalized the crack path observed in the non-proportional loadings. Doquet et al. [12] investigated through FEA the influence of static and 90° out-of-phase cyclic mode II loading superimposed to the cyclic mode I on the plasticity- and asperity-induced closure for an aluminum alloy. The applied mode II could either increase or decrease ΔK_{Ieff} depending on the stress ratio and loading path; in some cases, it had opposite effects on the two closure mechanisms. In the case of 90° out-of-phase mixed mode loading, mode II drastically reduced or even suppressed the plasticity-induced closure. To clarify the incipient direction of the crack path observed by Bold et al. [8], Dahlin and Olsson [13] investigated several criteria while using the FEA approach; they concluded that a maximum tangential stress (MTS) range criterion based on an elasto–plastic stress field can appropriately predict the incipient crack growth direction in ductile metals. Yu et al. [14] studied the FCG behavior in a thin-walled tubular specimen of aluminum alloy under non-proportional mixed mode I and II loading, showing that the FCG significantly differed from that under mode I or proportional mixed mode loading. Although they obtained a long and stable shear mode growth, they could not apply the commonly accepted MTS criterion in most of the non-proportional loading cases.

As indicated above, previous studies on coplanar crack growth under non-proportional mixed mode I/II loading cycles apparently focused on rail steel. Only Akama and Susuki [10] considered also the wheel steel case, but their elucidation of the phenomena associated with crack growth under such loading cycles was not sufficient. In particular, the reason why the growth direction of the crack changes depending on the loading conditions and the type of steel was not clarified. Besides, the crack growth models by Wong et al. [9] have been pointed out as arbitrary. Therefore, in this study, crack growth data from non-proportional mode I/II loading cycles were re-constructed by using a reliable equivalent SIF range and FEA was performed to elucidate the crack growth behavior under these loading cycles. The reliable criteria for predicting the crack path direction of non-proportional mixed mode loading were used, and the results were compared with corresponding experimental results.

This paper, part 1 of two companion papers, presents FCG under non-proportional mixed mode I/II loading and is organized as follows. The introduction starts in this section with a brief overview of past papers for the crack growth under non-proportional mixed mode I/II loading cycles. Section 2 describes the detailed method of the experiments conducted and presents the experimental results. In Section 3, the FEA model for predicting the crack path direction is presented and the results are indicated. Section 4 gives detailed considerations and discussions by comparing the experimental and FEA results. Finally, the important results obtained in this study are summarized in Section 5.

2. Experiments

2.1. Testing Machine

The crack growth under mixed mode I/II loading was investigated with an in-plane biaxial testing machine consisting of four hydraulic actuators, each one having a 200 kN tension–compression capacity for static and fatigue loads. The actuator assemblies formed two pairs of X and Y axes and were rigidly and diagonally mounted in an octagonal box-shaped frame, which was of heavy-duty welded construction and placed horizontally. A load cell and a linear variable differential transformer were installed in each actuator assembly. The components also included manifolds, a control console, and a hydraulic pump unit. To obtain and maintain the designated stress ratios during the fatigue tests, an improved control system was developed specifically for this study. The two actuator pairs were controlled by both load and positioning feedback signals. The control system of the load signals X1 and X2 for the *x*-axis actuator pair is schematized in Figure 2. The load command signal FX was derived from the load feedback signals FX1 and FX2 and from the position feedback signals S1 and S2, which were calculated, correspondingly, from F1 and F2 and from the calculation circuits CX1 and CX2; the notations F, S, 1, and 2 indicate, respectively, the load cell, the differential transformer output, and the X1 and X2 actuators. The same method was applied for the *y*-axis actuator. Testing frequencies from 1 to 2 Hz were used.

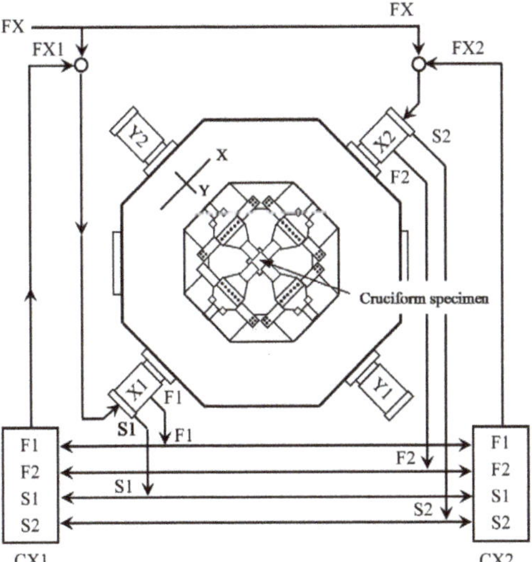

Figure 2. Schematic of the in-plane biaxial testing machine and the system controlling the load signals for the actuators.

2.2. Specimens

Cruciform specimens (Figure 3) having the 45° starter notch with a half-length of 2 mm, made by spark erosion, were used in the experiments [15]. Each specimen had a uniformly stressed square working section (72 × 72 mm, 4 mm in thickness) to observe crack growth behaviors. The tensile and compressive loads were applied, respectively, through pins and at the specimen edge; the backlash was removed via wedge tightening. The starter notch was pre-cracked by using an equibiaxial mode I loading with a zero stress ratio. The load was reduced during pre-cracking in 10% steps until an FCG rate of less than 10^{-9} m/cycle was reached. This was to reduce the residual plastic zone size to less than that produced during the first cycle of the test.

Two-rail steels (RP and RF) and one-wheel steel (WT) were used as specimen materials. The RP and RF were used as normal and head hardened rails, respectively, that were installed for corresponding straight sections and outer rails in curved sections. Their chemical compositions and mechanical properties are summarized in Tables 1 and 2, respectively, showing that the ultimate tensile strength and 0.2% proof stress of RF were much superior to those of RP and WT. The microstructures of RP and WT were normal pearlite, whereas that of RF was fine pearlite with an average lamellar spacing of about 100 nm. The specimens were collected directly from real rails and wheels.

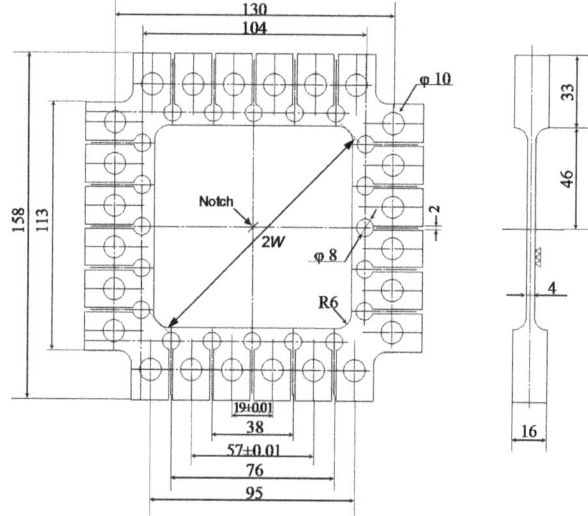

(a) Whole configuration (all dimensions in mm)

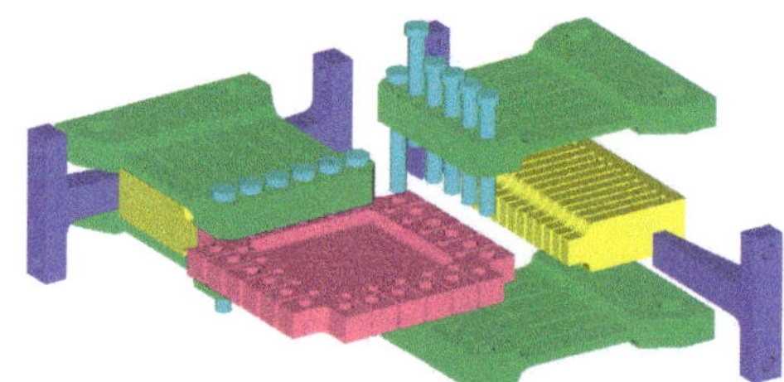

(b) Arrangement of specimen, tensile and compressive grips, pin and wedges

Figure 3. (a) Configuration of the cruciform specimen used in the experiments (all dimensions in mm) and (b) its arrangement with the tensile and compressive grips, pin, and wedges.

Table 1. Chemical composition of steels used for the experiments (wt%).

Material	C	Si	Mn	P	S
Rail steel, RP	0.68	0.26	0.93	0.016	0.01
Rail steel, RF	0.79	0.17	0.82	0.019	0.01
Wheel steel, WT	0.65	0.26	0.73	0.016	0.01

Table 2. Mechanical properties.

Material	Ultimate Tensile Strength (MPa)	0.2% Proof Stress (MPa)
Rail steel, RP	934	511
Rail steel, RF	1214	802
Wheel steel, WT	from 981 to 1030	from 618 to 657

2.3. Loading History

The loading history, illustrated in Figure 4, simulated that experienced by RCF cracks in the presence of a fluid, as obtained by FEA [2–5]. Since the load is represented by sine waves, the time on the horizontal axis is indicated as an angle (degree); the figure also shows δ between K_I and K_{II}. When a contact area between the wheel and the rail approaches a surface crack, the surface tangential traction slightly opens the crack mouth in the driven surface and the surrounding fluid penetrates inside it. When the contact area moves over the crack mouth, the K_I load increases (0–180°) due to the fluid entrapment [16] or the hydraulic pressure mechanism [17]; after it leaves the crack mouth, the fluid in it flows out and its influence on K_I almost disappears (180°). Meanwhile, the K_{II} loading is also generated and completely reversed (90–450°). Hence, if there is fluid inside the crack, cracks in the triaxial compressive stress field can open in the contact area and large K_I and K_{II} loading are generated.

This condition was simulated by actuating both axes of the biaxial machine to apply the required mode I and II loading sequence to the crack. Figure 5 shows loading examples of the x- and y-axes for a 45° crack.

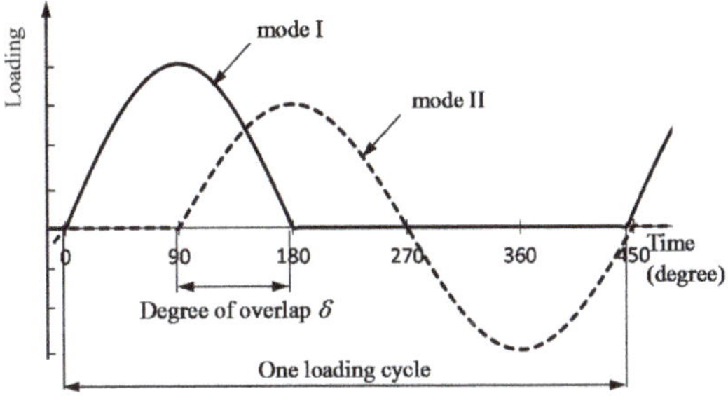

Figure 4. Loading cycle applied under the condition of $\Delta K_{II}/\Delta K_I = 1.5$ and $\delta = 90°$.

2.4. Calculation of the Stress Intensity Factors

K_I and K_{II} were calculated as

$$K_I = Y\sigma\sqrt{\pi a} \tag{1}$$

$$K_{II} = Y\tau\sqrt{\pi a} \tag{2}$$

where $Y = \sqrt{\sec(\pi a/2W)}$ for the central 45° crack in a plate of diagonal length $2W$ (refer to Figure 3) and σ and τ are the normal and in-plane shear stress, respectively, applied to the crack. The crack length (a) was measured with a traveling microscope at the resolution of 1 μm. In addition, ΔK_{Ieff} and ΔK_{IIeff} were defined by the ratios of crack closure (U_I)and locking (U_{II}), respectively [18]

$$\Delta K_{\text{Ieff}} = U_{\text{I}} \Delta K_{\text{I}} \tag{3}$$

$$\Delta K_{\text{IIeff}} = U_{\text{II}} \Delta K_{\text{II}} \tag{4}$$

$$U_{\text{I}} = \frac{v_{a\max} - v_{a\min}}{v_{th\max} - v_{th\min}} \tag{5}$$

$$U_{\text{II}} = \frac{u_{a\max} - u_{a\min}}{u_{th\max} - u_{th\min}} \tag{6}$$

where $v_{a\max}$, $v_{a\min}$, $v_{th\max}$, and $v_{th\min}$ are the measured maximum, measured minimum, theoretical maximum, and theoretical minimum opening displacements, respectively, and $u_{a\max}$, $u_{a\min}$, $u_{th\max}$, and $u_{th\min}$ indicate the corresponding values for the sliding displacement. The crack closure and locking were measured via the surface replica technique.

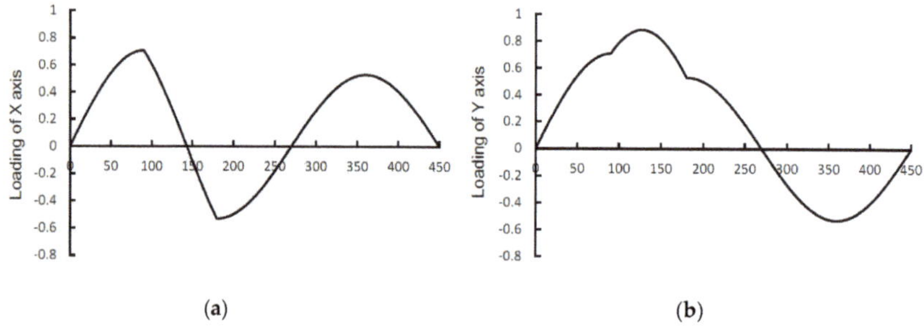

(a)   (b)

Figure 5. Sequential loading cycle of the (**a**) x and (**b**) y axes under the condition of $\Delta K_{\text{II}}/\Delta K_{\text{I}} = 1.5$ and $\delta = 90°$; the vertical axis represents the relative values of the loads of x- and y-axes.

2.5. Experimental Conditions

The experiments were performed on four WT, six RP, and one RF specimen and, hereafter, will be accordingly referred to as WT1, WT2, WT3, and WT4; RP1, RP2, RP3, RP4, RP5, and RP6; and RF1; respectively. The stress ratios of mode I and mode II loading (R_{I} and R_{II}) were 0 and –1, respectively. Table 3 shows the $\Delta K_{\text{II}}/\Delta K_{\text{I}}$ ratios and δ values for each experiment; WT1–WT4; RP1–RP4, and RF1 were designed to mainly produce coplanar crack growth rate data, whereas RP5 and RP6 were conducted for obtaining branch crack growth rate data.

Table 3. Testing conditions.

Exp. No.	$\Delta K_{\text{II}}/\Delta K_{\text{I}}$	δ(degree)
WT1	1.0	0, 10, 20, 30, 40, 50, 60, 70, 80, 90, 120
WT2	1.0	30, 60, 90, 120
WT3	1.25	0, 10, 20, 30, 40, 50, 60, 70, 80, 90, 120
WT4	1.0, 1.375, 1.5, 1.9, 2.0	10, 30, 60
RP1	1.0	0, 10, 20, 30, 40, 50, 60, 70, 80, 90, 120
RP2	1.5	0, 10, 20, 30, 40, 50, 60, 70, 80, 90
RP3	1.0	30, 60, 90, 120, 150
RP4	1.5	30, 60, 90, 120
RP5	2.0	0
RP6	2.5	0
RF1	1.0, 1.375, 1.5	30

2.6. Experimental Results

A polar coordinate system having the origin coinciding with the crack tip was introduced, as shown in Figure 6 together with a schematic representation of the main coplanar crack plane and its angle (θ) with the branch crack plane, when branching occurs.

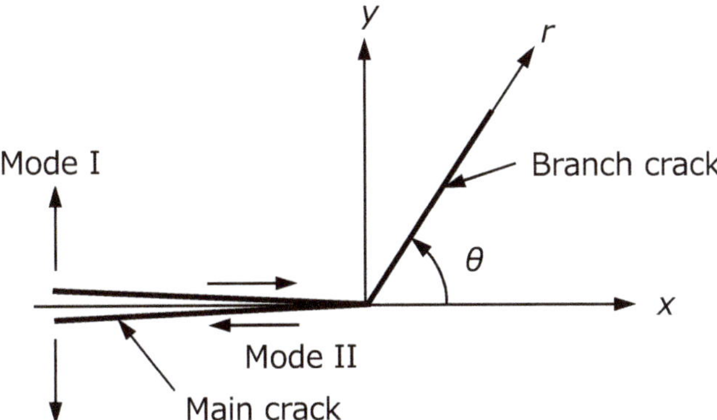

Figure 6. Branch crack at a main crack plane and branch angle definition.

The coplanar crack grew almost straight (i.e., $\theta = 0$) for all δ values in WT1–WT3, RP1–RP3, and RF1. After the crack length on each side extended to about 3 mm, δ increased during 10° or 30° steps. The ΔK_I and ΔK_{II} values were maintained approximately constant for each loading step by decreasing the loads. In WT4, the coplanar crack grew when $\Delta K_{II}/\Delta K_I = 1.0$ regardless of δ, but it branched when this ratio exceeded 1.375. Immediately after this branch crack growth, the crack path was restored to an initial 45° plane by using an equibiaxial mode I loading and, then, the experiments were continued. In RP4, the coplanar crack branched at a θ of about –70° when δ was 120°. When the crack was short, U_I and U_{II} at the crack tip were hard to estimate. However, some v_{amax}, v_{amin}, u_{amax}, and u_{amin} measurements were performed near the crack tip and the ratios U_I and U_{II} were estimated by extrapolating back to the crack tip. In RP5 and RP6, branch cracks grew; for these branched cracks, the displacements to obtain U_I and U_{II} were not measured.

2.6.1. Coplanar Crack Growth Rate

The coplanar crack growth rates of RP, which comprises RP1, RP2, and RP3 obtained from the mixed mode I/II tests, were plotted against ΔK_I (Figure 7), including all δ cases and, for comparison, the pure mode I growth data collected at $R_I = 0$. The FCG rates for the mixed mode I/II loading were fairly faster than those for pure mode I. When the Paris-type law was applied to the crack growth rates for all the mixed mode I/II loading data, we obtained the equation

$$\frac{da}{dN} = 3.75 \times 10^{-12} (\Delta K_I)^{3.39} \tag{7}$$

where N is the number of cycles. The coefficient of determination (R^2) was 0.69.

Next, the same growth rates were plotted against ΔK_{II} (Figure 8), revealing an improvement in R^2 (0.87)

$$\frac{da}{dN} = 4.80 \times 10^{-12} (\Delta K_{II})^{3.12} \tag{8}$$

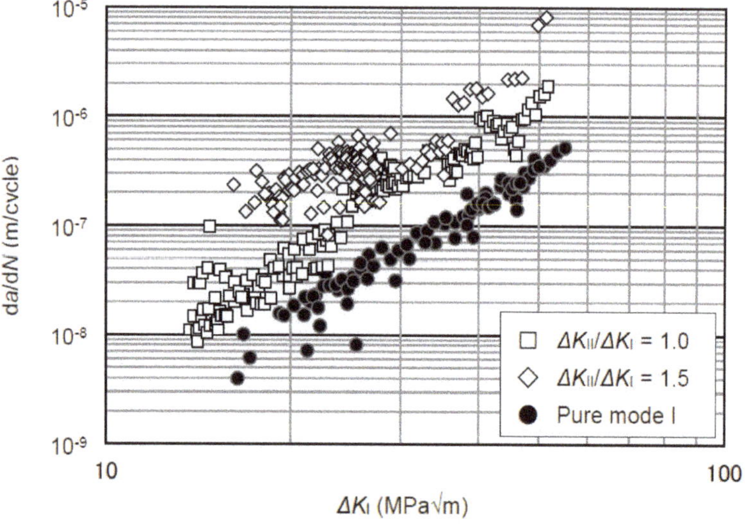

Figure 7. Coplanar crack growth rates against ΔK_I comparing pure mode I data for RP.

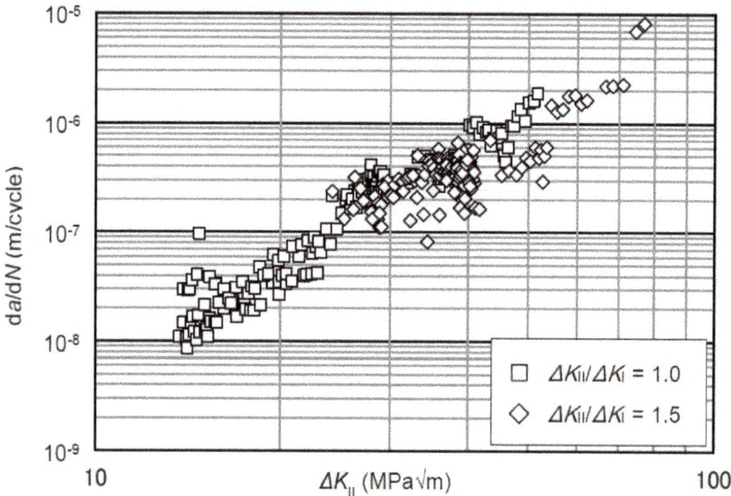

Figure 8. Coplanar crack growth rates against ΔK_{II} for RP.

Then, since the crack growth law can be described uniquely even under different loads and stress ratios by effective SIFs, the growth rates were plotted against ΔK_{Ieff} (Figure 9). The crack growth rate data can be summarized by the equation

$$\frac{da}{dN} = 8.19 \times 10^{-11} (\Delta K_{Ieff})^{2.59}$$

(9)

In this case, a further correlation improvement was obtained ($R^2 = 0.89$). The ΔK_{Ieff} and ΔK_{IIeff} values were successively used to fit the growth data.

The crack growth rates were then plotted against the equivalent SIF range (ΔK_v), as proposed by Richard et al. [19] (Figure 10), which is defined as

$$\Delta K_v = \frac{\Delta K_I}{2} + \frac{1}{2}\sqrt{\Delta K_I^2 + 4(1.115\Delta K_{II})^2} \tag{10}$$

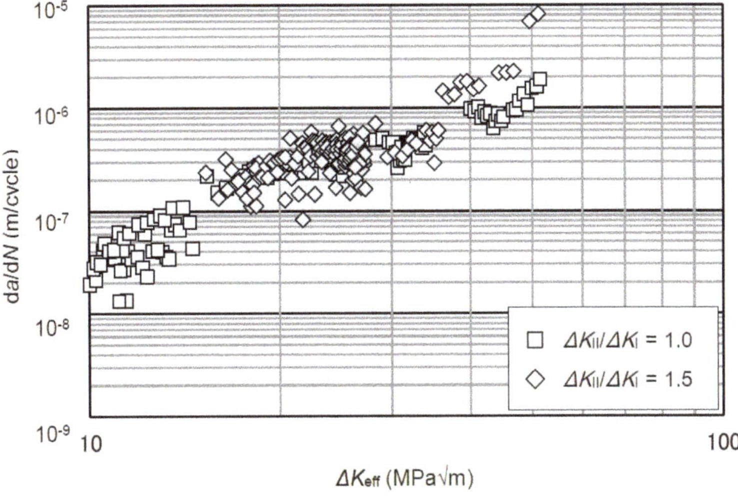

Figure 9. Coplanar crack growth rates against ΔK_{Ieff} for RP.

The crack growth rates were expressed as

$$\frac{da}{dN} = C(\Delta K_v)^m \tag{11}$$

In the RP, RF, and WT cases, the corresponding C and m values were 6.08×10^{-11} and 2.18, 8.70×10^{-11} and 2.04, and 3.82×10^{-11} and 2.45, respectively. Figure 10 shows that better correlations were attained when plotting the crack growth rates against ΔK_v. Over all, WT exhibited the fastest crack growth rates and, among the rail steels, the rates of RP were higher than those of RF.

2.6.2. Branch Crack Growth Rate

In RP, branch cracks occurred and grew where $\Delta K_{II}/\Delta K_I$ exceeded 2.0 and δ was 0°. For this specimen geometry, the branch crack was subjected to a large mode I cycle, which was generated from the mode II loading, followed by a small mode I cycle from the mode I loading. Therefore, the growth rates were plotted against the ΔK_I values from the mode II loading cycles only; ΔK_I was calculated using the procedure proposed by Gao et al. [20], which considers the stresses acting normal to the crack plane. In particular, the branch crack growth rates of RP were plotted against ΔK_I (Figure 11). These rates were expressed as

$$\frac{da}{dN} = 1.11 \times 10^{-11}(\Delta K_I)^{2.40} \tag{12}$$

for RP5 and as

$$\frac{da}{dN} = 2.64 \times 10^{-10}(\Delta K_{\mathrm{I}})^{1.95} \tag{13}$$

for RP6. As can be seen, the crack growth rate differs considerably depending on $\Delta K_{\mathrm{II}}/\Delta K_{\mathrm{I}}$, and increasing in the $\Delta K_{\mathrm{II}}/\Delta K_{\mathrm{I}}$ increases the crack growth rate.

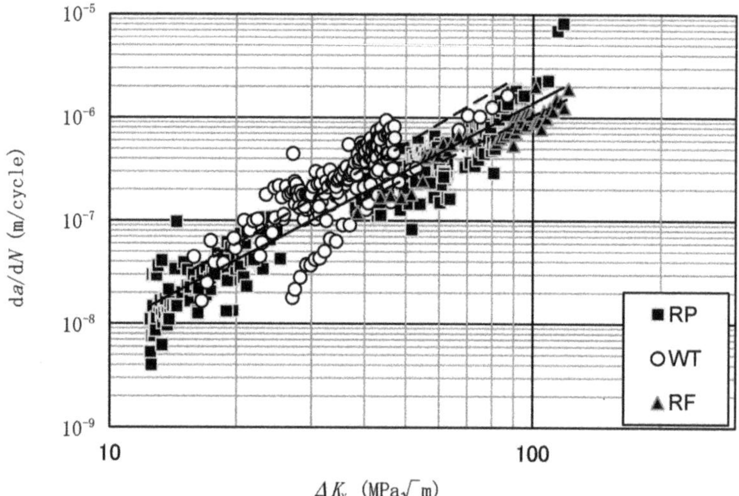

Figure 10. Coplanar crack growth rates of mixed mode I/II against ΔK_{v} for RP, RF and WT (solid line; RP, one dot chain line; RF, broken line; WT).

2.6.3. Fractography

To understand the fracture mechanism, scanning electric microscope (SEM) observations were performed. Figure 12 shows the fracture surfaces near the crack tip region and along the crack flank resulting from RP2, WT3, and RF1. Heavily deformed ridges and valleys were rubbed out in the sliding direction and also several oxidized wear particles were found on the surface. No striation pattern was observed near the crack tip region in case of WT3 and RF1, but RP2 exhibited obscure striations that seemed to be scattered. Hence, the fractographic observation permitted the determination of the crack growth caused by non-proportional mixed mode I/II loading, which was clearly distinguishable from that generated by pure mode I loading.

In summary, these were the crack growth characteristics observed in rail and wheel steel specimens subjected to non-proportional mixed mode I/II loading:

(1)   When $\Delta K_{\mathrm{II}}/\Delta K_{\mathrm{I}}$ increased, the crack tended to branch;
(2)   As δ increased, the crack easily branched;
(3)   The coplanar crack growth rates in RF were lower than those in WT and RP;
(4)   The branch crack growth rate varied considerably depending on $\Delta K_{\mathrm{II}}/\Delta K_{\mathrm{I}}$ even in the same material, unlike the coplanar growth rate, could not be correlated by a single line;
(5)   No clear striation patterns were found near the crack tip region.

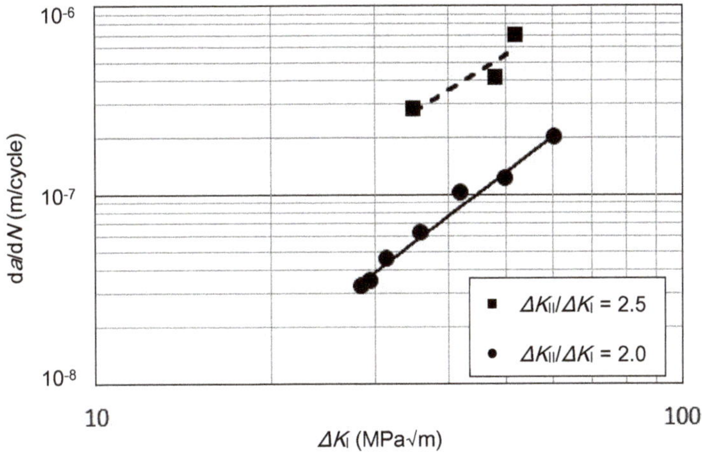

Figure 11. Branch crack growth rates of mixed mode I/II against ΔK_I for RP.

3. Finite Element Analysis

3.1. Procedure

The elasto–plastic analysis was performed using the commercial FEA code MARC to better elucidate the crack growth characteristics. A stationary crack was considered because, in this study, a branch angle θ from the main crack was predicted. Figure 13 shows the mesh used, with a crack length a fixed at 5 mm; only the uniformly stressed square working section was considered, represented by a simple square in plain strain assumption with a thickness of 4 mm. Second-order isoparametric quadrilateral and triangular elements, refined to 25 μm near the crack tip, were used with the singular elements [21] arranged around the tips; for smaller elements, they were greatly distorted due to a large number of loading cycles, resulting in remarkably poor analysis accuracy. The total numbers of elements and nodes were 644 and 1968, respectively. The boundary conditions applied for cruciform specimens under mixed mode I/II loading were also described. At each edge center, a pilot node was connected to the remaining nodes of that edge through a multipoint constraint; all the loads were applied to these pilot nodes as concentrated normal and shear forces (F and S, respectively). The use of this mesh and loading procedure allowed the confirmation that the MTS occurred at $\theta = -70.5°$ due to pure mode II loading in the elastic analysis. The contact between crack faces was taken into account and the friction coefficient was set to 0.3.

When analyzing the fatigue problems, the choice of the material constitutive model is crucial. In this study, the model combining the nonlinear kinematic hardening rule with the isotropic hardening rule, developed by Chaboche and Lemaitre [22] (C&L model), was used as

$$^{ti+\Delta ti}\sigma_y = {}^{0}\sigma_y + Q\left\{1 - \exp\left(-B^{ti+\Delta ti}\bar{e}^p\right)\right\} \tag{14}$$

$$\mathrm{d}\alpha = \frac{2}{3}h\mathrm{d}e^p - \zeta\alpha\mathrm{d}\bar{e}^p \tag{15}$$

where $^{ti+\Delta ti}\sigma_y$ is the updated yield stress at time $t_i + \Delta t_i$, $^{0}\sigma_y$ is the initial yield stress, Q, B, h, and ζ are material constants, $^{ti+\Delta ti}\bar{e}^p$ is the accumulated effective plastic strain at $t_i + \Delta t_i$, α is the shift of the yield surface center, e^p is the plastic strain, and d of $\mathrm{d}\alpha$ and $\mathrm{d}\bar{e}^p$ implies increment. The material constants for RP and RF, determined via strain controlled uniaxial fatigue experiments, are summarized in Table 4 along with Young's modulus E and Poisson ratio ν. The FEA was performed on these two rail steels to clarify the material effect on the crack growth rate.

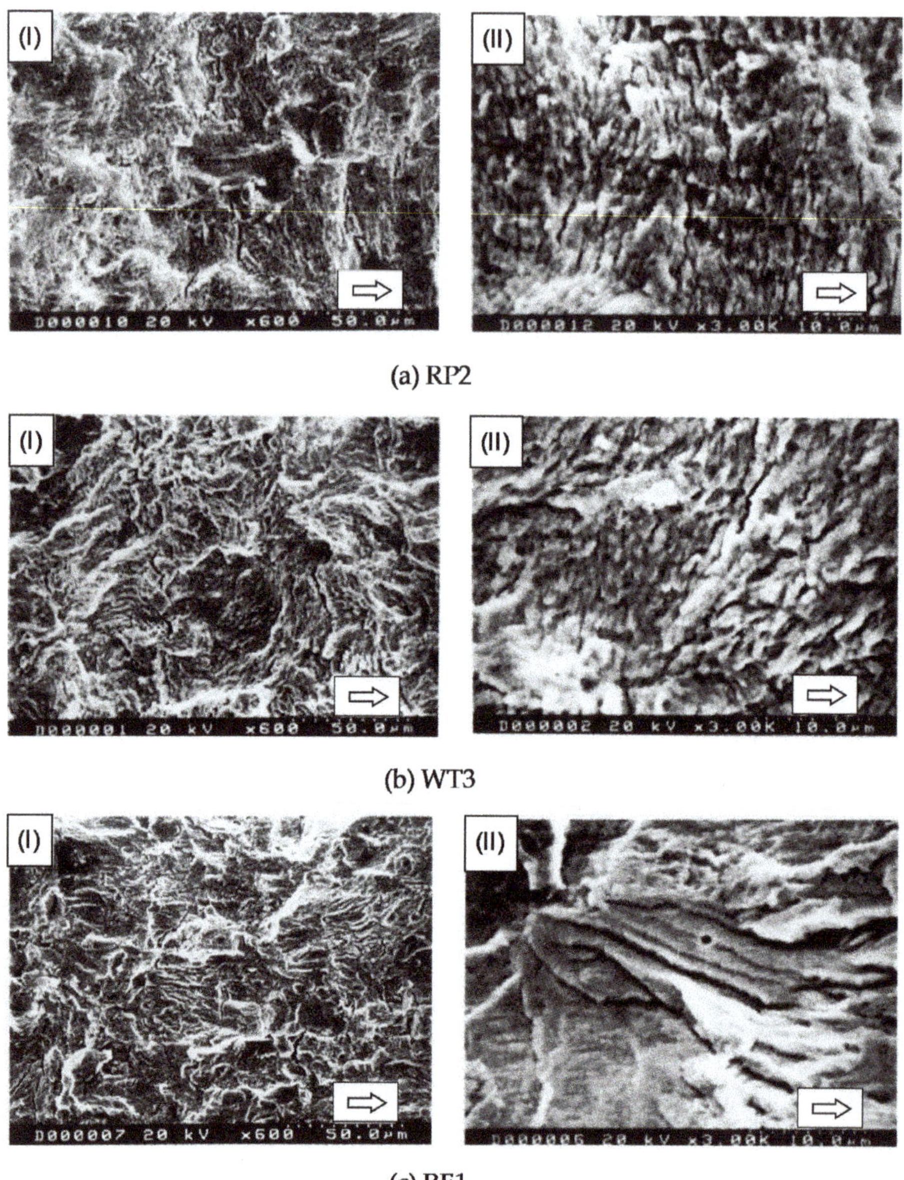

(a) RP2

(b) WT3

(c) RF1

Figure 12. Fracture surface near the crack tip; (**a**) RP2, (**b**) WT3, and (**c**) RF1. Arrows indicate the direction of crack growth.

3.2. Analytical Results

The analytical conditions (Table 5) were decided based on the actual experiments described in Section 2. Figures 14 and 15 show some results about the distribution of the tangential (σ_θ) and shear stresses ($\tau_{r\theta}$), respectively, in the first loading cycle for ARP2; the maximum $\Delta\sigma_\theta$ appeared between $-45°$ and $-67.5°$, whereas the maximum $\Delta\tau_{r\theta}$ was observed exactly at $0°$.

When using the C&L model, several loading cycles are required to achieve a steady-state stress cycle; therefore, at least 100 loading cycles were initially planned for the simulation. However, as the number of cycles increased, the deformation of the crack tip elements increased accordingly and, hence, an accurate analysis was considered no longer possible after the 50th cycle. Thus, the evaluations were performed at the 50th loading cycle and, at this point, a steady-state stress cycle could not be achieved.

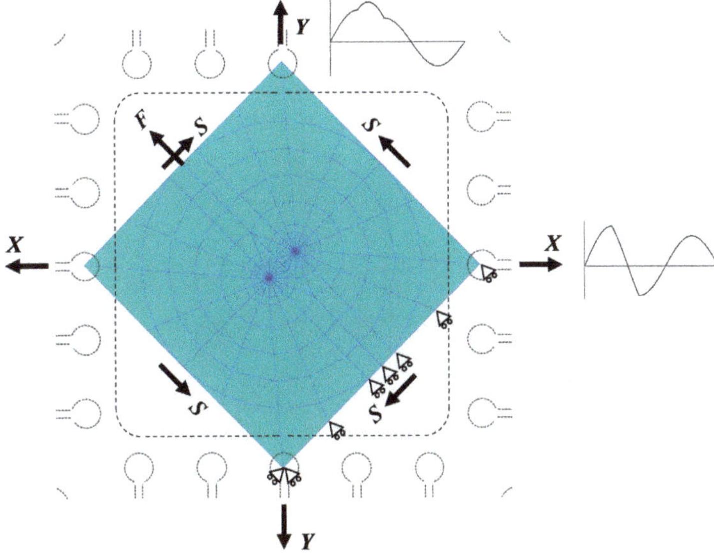

Figure 13. Finite element mesh and boundary conditions for the cruciform specimen under mixed mode I/II loading.

Table 4. Material properties used in FEA.

	E (MPa)	ν	$^0\sigma_y$ (MPa)	Q	b	h (MPa)	ζ
RP	183,008	0.3	508	−208	24.2	85,248	193
RF	182,778	0.3	684	−264	1.27	88,615	185

Table 5. FEA conditions.

No.	$\Delta K_{II}/\Delta K_{I}$	δ (degree)	F (N)	S (N)	Material
ARP1	1.5	90	$0 \leftrightarrow 26{,}667$	$-20{,}000 \leftrightarrow 20{,}000$	RP
ARP2	1.5	120	$0 \leftrightarrow 26{,}667$	$-20{,}000 \leftrightarrow 20{,}000$	RP
ARP3	1.5	30	$0 \leftrightarrow 26{,}667$	$-20{,}000 \leftrightarrow 20{,}000$	RP
ARF1	1.5	30	$0 \leftrightarrow 26{,}667$	$-20{,}000 \leftrightarrow 20{,}000$	RF

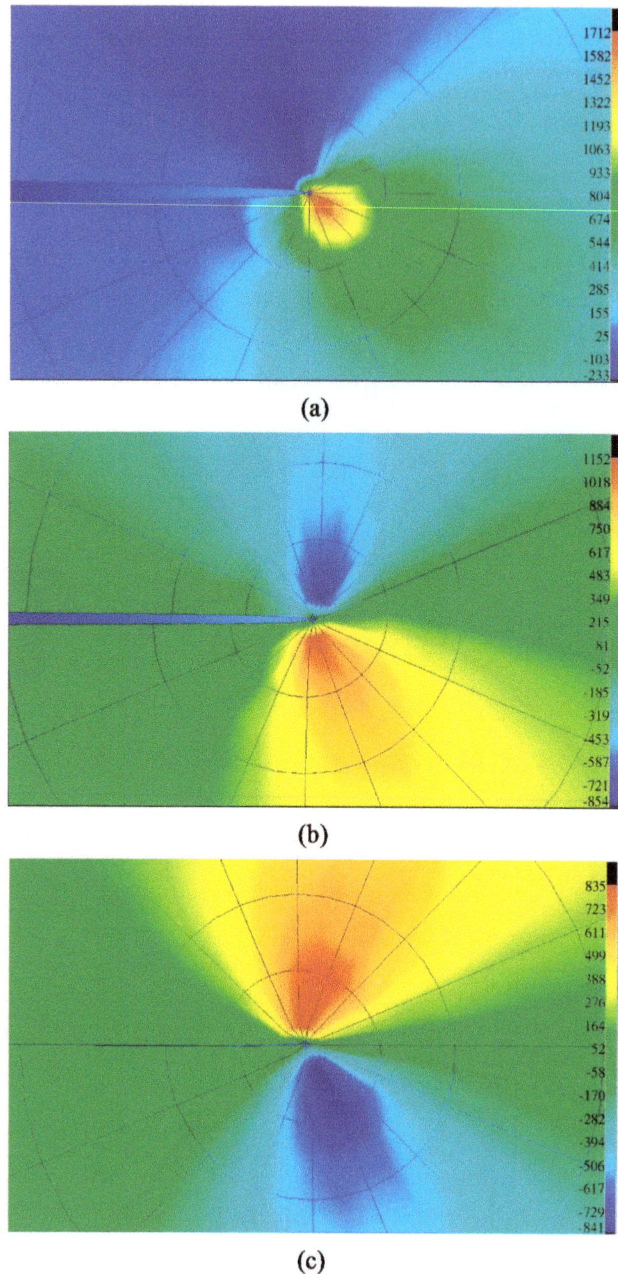

Figure 14. Tangential stress distribution near the crack tip for ARP2 ($\Delta K_{II}/\Delta K_I = 1.5$, $\delta = 120°$) at the (**a**) 90° (maximum K_I), (**b**) 150° (maximum K_{II}), and (**c**) 330° (minimum K_{II}); contour levels in MPa and the deformation magnification is 5×.

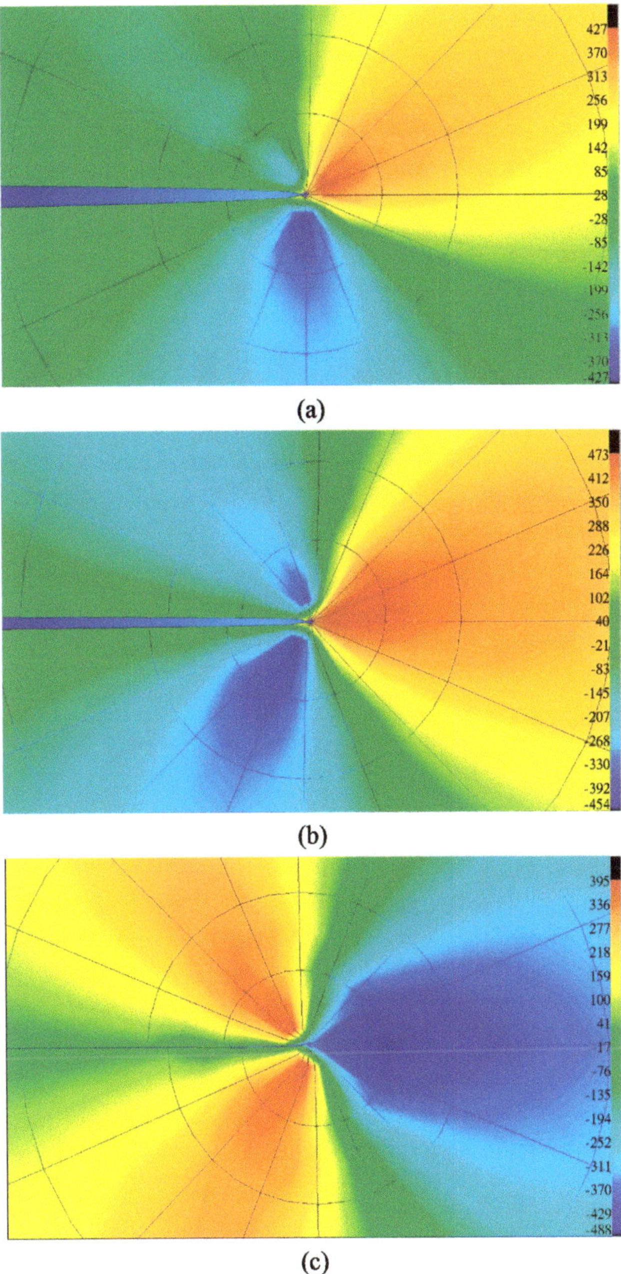

Figure 15. Shear stress distribution near the crack tip for ARP2 ($\Delta K_{II}/\Delta K_{I} = 1.5$, $\delta = 120°$) at the (**a**) 90° (maximum K_{I}), (**b**) 150° (maximum K_{II}), and (**c**) 330° (minimum K_{II}); contour levels in MPa and the deformation magnification is 5×.

3.2.1. Angles of Maximum Normal and Shear Stress Ranges

In this section, the purpose was to predict the incipient angle of crack growth after a change in the applied load from the condition, which leads the crack to stable coplanar growth. Dahlin and Olsson [13] suggested that the MTS range based on an elasto–plastic stress field could capture the effects of $\Delta K_{II}/\Delta K_I$ and δ on crack branching in rail steel. According to this criterion, the crack deviation angle lies along the material plane on which the tangential stress range ($\Delta\sigma_\theta$) is maximum during the selected load cycle. In this study, the $\Delta\sigma_\theta$ level on every plane was investigated; when σ_θ was below zero, it was reset to zero in the calculation. The maximum $\Delta\sigma_\theta$ value and the plane on which it appeared were indicated, respectively, as $\Delta\sigma_{max}$ and $\theta_{\sigma max}$. The shear stress range ($\Delta\tau_{r\theta}$) on every plane was also investigated. This underlying assumption was adopted: if the cracks grow via a shear mode, the growth direction should be determined by the plane ($\theta_{\tau max}$) on which $\Delta\tau_{r\theta}$ became maximum ($\Delta\tau_{max}$).

As mentioned in Section 2, $\Delta K_{II}/\Delta K_I$ and δ could influence the crack growth direction. More specifically, when $\Delta K_{II}/\Delta K_I$ increased from 1 to 1.5 with $\delta = 120°$, the crack growth changed from coplanar into branch; a δ increase from 90° to 120° while maintaining $\Delta K_{II}/\Delta K_I = 1.5$ changed the coplanar growth into branch growth; this was explicitly clarified by Dahlin and Olsson, namely, if this ratio was increased, the $\Delta\sigma_\theta$ direction switched from $\theta = 0°$ to $\theta = -70°$. Therefore, in this study, the δ influence on the crack growth direction was clarified.

The FEA results about the values of $\Delta\sigma_{max}/\Delta\tau_{max}$, $\theta_{\sigma max}$, and $\theta_{\tau max}$, when changing δ and maintaining $\Delta K_{II}/\Delta K_I = 1.5$, which is determined by comparing ARP1 and ARP2, are summarized in Table 6; when δ increased, $\Delta\sigma_{max}/\Delta\tau_{max}$ increased, whereas $\theta_{\sigma max}$ and $\theta_{\tau max}$ remained approximately on the branch and coplanar directions, respectively.

Table 6. Effect of δ on $\Delta\sigma_{max}/\Delta\tau_{max}$, $\theta_{\sigma max}$, and $\theta_{\tau max}$.

No.	$\Delta K_{II}/\Delta K_I$	δ (degree)	$\Delta\sigma_{max}/\Delta\tau_{max}$	$\theta_{\sigma max}$ (degree)	$\theta_{\tau max}$ (degree)
ARP1	1.5	90	1.45	−70	0
ARP2	1.5	120	1.56	−78	0

3.2.2. Crack Tip Opening and Sliding Displacements

The crack tip displacements are important since they could influence the crack growth rate and path direction. Therefore, the FEA results were used to obtain the crack tip opening displacement (CTOD), the crack tip sliding displacement (CTSD), and the residual crack tip opening displacement, which corresponded to the CTOD when the 50th loading was ended, of RP and RF. Figure 16 compares the resulting ranges of CTOD (ΔCTOD), CTSD (ΔCTSD), and residual CTOD of the two different materials (such as the comparison of ARP3 and ARF1). All these values were smaller for the RF case.

4. Discussion

The coplanar growth rates were plotted against ΔK_I, ΔK_{II}, and ΔK_{Ieff} for RP; the effective values were obtained considering that the crack closure and locking ratios were changed with respect to δ. ΔK_{II} gave satisfactory correlations compared to ΔK_I, suggesting that the crack growth rates were more influenced by mode II loading than by mode I loading. Moreover, ΔK_{Ieff} also provided relatively satisfactory correlations; however, these correlations were not sufficient because the crack growth contributions of mode I and II loading cycles were not mutually independent and, hence, the single mode range alone could not represent the crack growth rates uniquely. Therefore, the crack growth rates were plotted against ΔK_v, which considered the interaction of the two loading cycles; although ΔK_v was invented for proportional loading, it gave satisfactory correlations.

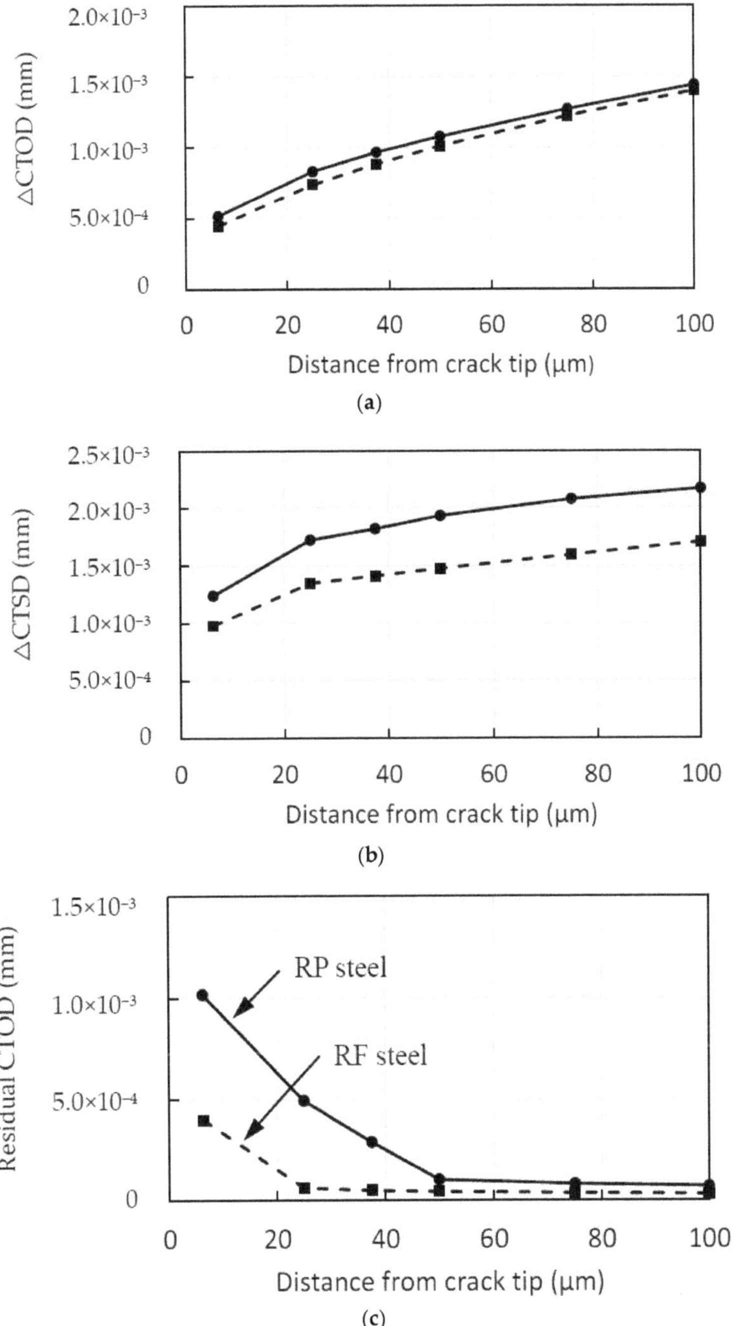

Figure 16. Ranges of (**a**) crack tip opening and (**b**) sliding and (**c**) residual crack tip opening displacement of RP (solid line) and RF (dashed line). ($\Delta K_{II}/\Delta K_I = 1.5$, $\delta = 30°$).

The branch crack growth rates could not be represented by a single line, as shown in Figure 11. All the cruciform specimens were tested under biaxial loading conditions and, in these cases, the non-singular stress called the *T*-stress acted at the crack tips. If the crack angle is not 45°, as in the branch crack case, the *T*-stress affects the plasticity near the crack tip and the plastic zone size increases with the shear part of the loading. Because a larger plastic zone yields higher growth rates [19], a separation of the data for $\Delta K_{II}/\Delta K_I = 2.5$ and $\Delta K_{II}/\Delta K_I = 2.0$ was observed, i.e., when the rate of shear loading was higher, the growth rates became higher.

The fractographic observation by SEM showed no clear striation patterns on the fracture surfaces near the crack tips, while roughness due to friction was observed in the case of RF1 and WT3. Such surface damage usually arises from the interaction between the crack faces under shear mode loading, which was reported in Fujii et al. [23]; therefore, the main crack growth mechanism observed in this study was assumed as caused by mode II loading.

The experimental results suggested that a δ increase could encourage crack branching. This can be elucidated by FEA; when δ increased, $\Delta\sigma_{max}/\Delta\tau_{max}$ increased and the $\Delta\sigma_{max}$ plane was oriented toward the branch direction, whereas the $\Delta\tau_{max}$ plane remained on the coplanar plane. The widely accepted criterion to predict the crack growth direction under non-proportional loading states that the crack selectively grows along the fastest growth direction. Based on this criterion, the FEA results can explain the experiment results.

The stresses were evaluated at the center of the elements around the crack tip, whose size was 25 μm. Because the MTS range criterion based on an elasto–plastic stress field is considered, these elements should be included inside the plastic zone ahead of the crack tip. The plastic-zone size under the investigated mixed mode I/II loading was not clear. Here, the cyclic crack tip plastic zone size (r_p) was derived from the analytical solutions based on pure mode I and mode II loading. Under the FEA conditions adopted, the size developed by mode II loading was larger than that resulting from mode I. If the size extension due to the stress redistribution is not taken into account, the size developed by mode II loading can be roughly estimated for both plane stress and strain conditions as

$$r_p = \left(\frac{1}{2\pi}\right)\left(\frac{\Delta K_{II}}{2\tau_{ys}}\right)^2 \tag{16}$$

where τ_{ys} is the yield stress of the material under shear. Based on all the conditions analyzed using RP, which were $\Delta K_{II} = 16.6$ MPa√m and $\tau_{ys} = 293$ MPa, the r_p was calculated as 128 μm. Therefore, the points at which the stresses were evaluated were still far inside the plastic zone.

The stress evaluation was performed in the 50th loading cycle, although the steady-state stress distribution was not achieved because the distortion of elements increased, deteriorating the FEA accuracy. MARC provides a function to subdivide the mesh when the element distortion is large; one of the criteria is the maximum change in an interior angle from the initial angle for triangle elements and the recommended angle is 40°. For the ARP2 case, this criterion was violated at the 100th cycle and, hence, the stresses were evaluated at the 50th cycle with a margin.

However, remeshing was not performed in this study; the crack growth was expected to occur during the loading periods under the present testing conditions. In the ARP2 case, for example, ΔK_I and ΔK_{II} were 11.1 MPa√m and 16.6 MPa√m, respectively. Based on the crack growth law expressed in Equations (10) and (11), the crack growth length in 50 loading cycles was 3.4 μm, and the stress redistribution near the crack tip occurred according to this crack extension.

ΔCTOD and ΔCTSD were obtained for both RP and RF via FEA since they can influence the crack growth rate. In fact, a vector crack tip displacement (CTD) criterion has been proposed by Li [24] to correlate mixed mode FCG rates. This criterion was based on the concept that the vector CTD range (ΔCTD) is the driving force for FCG. Here, the vector ΔCTD corresponded to the vector summation of ΔCTOD and ΔCTSD. The FEA results indicated that ΔCTOD and ΔCTSD of RF were smaller than those of RP, probably because the former was a high-strength steel (Table 2); besides, RF exhibited much smaller residual CTOD and its opening distance under no loading compared to RP. Since there were irregularities on the crack faces, there is a possibility that the mode II loading might have been attenuated, especially in RF. This implies that the crack growth rate in RF was lower than that in RP, explaining the experimental results.

5. Conclusions

Fatigue tests were conducted on rail and wheel steel specimens to obtain FCG rates under mixed mode I/II loading cycles, which can simulate rolling contact conditions, by using an in-plane biaxial fatigue machine. Simplified cycles were applied to the analysis of RCF cracks, including the effect of fluid trapped inside the cracks. To elucidate the experimental results, a FEA was also performed. The main findings can be summarized as follows.

(1) When the stress intensity range ratio between mode II and I increased, the crack tended to branch. This happened because, when this ratio increased, the direction of the maximum tangential stress range changed from coplanar growth to branch one.

(2) As the degree of overlap between the mode I and II stress intensities increased, the crack easily branched; in particular, the MTS range became superior to the maximum shear stress range and its plane oriented toward the branch direction.

(3) Coplanar crack growth rates were lower in head hardened rail steel than in normal rail steel. This was due to the smaller CTDs in head hardened rail steel, which can influence the growth rate, compared to normal rail steel.

(4) The branch crack growth rate varied considerably depending on the stress intensity range ratio even within the same rail steel and, unlike the coplanar growth rate, could not be correlated by a single line because the plastic zone size increased with the shear part of the loading as a result of the T-stress and this larger plastic zone gave higher growth rates.

(5) Clear striation patterns were not found near the crack tip region. Therefore, the main driving force for the crack growth under the investigated non-proportional mixed mode I/II loading was thought to be the mode II loading.

Funding: This research received no external funding.

Acknowledgments: The author thank Ippei Susuki of the Japan Aerospace Exploration Agency (JAXA) for his support in developing the test machine.

Conflicts of Interest: The author declares no conflict of interest.

References

1. Grassie, S.L. Squats and squat-type defects in rails: The understanding to date. *Proc. IMechE Part F J. Rail Rapid Transit* **2012**, *226*, 235–242. [CrossRef]
2. Bogdanski, S.; Olzak, M.; Stupnicki, J. Numerical modeling of a 3D rail RCF 'squat'-type crack under operating load. *Fatigue Fract. Eng. Mater. Struct.* **1998**, *21*, 923–935. [CrossRef]
3. Akama, M.; Mori, T. Boundary element analysis of surface initiated rolling contact fatigue cracks in wheel/rail contact systems. *Wear* **2002**, *253*, 35–41. [CrossRef]
4. Bogdanski, S.; Lewicki, P. 3D model of liquid entrapment mechanism for rolling contact fatigue cracks in rails. *Wear* **2008**, *265*, 1356–1362. [CrossRef]
5. Akama, M.; Nagashima, T. Some comments on stress intensity factor calculation using different mechanisms and procedures for rolling contact fatigue cracks. *Proc. IMechE Part F J. Rail Rapid Transit* **2009**, *223*, 209–221. [CrossRef]
6. Fremy, F.; Pommier, S.; Poncelet, M.; Raka, B.; Galenne, E.; Courtin, S.; Roux, J.-C.L. Load path effect on fatigue crack propagation in I + II + III mixed mode conditions–Part 1: Experimental investigations. *Int. J. Fatigue* **2014**, *62*, 104–112. [CrossRef]
7. Bold, P.E.; Brown, M.W.; Allen, R.J. Shear mode crack growth and rolling contact fatigue. *Wear* **1991**, *144*, 307–317. [CrossRef]
8. Wong, S.L.; Bold, P.E.; Brown, M.W.; Allen, R.J. A branch criterion for shallow angled rolling contact fatigue cracks in rails. *Wear* **1996**, *191*, 45–53. [CrossRef]

9. Wong, S.L.; Bold, P.E.; Brown, M.W.; Allen, R.J. Fatigue crack growth rates under sequential mixed-mode I and II loading cycles. *Fatigue Fract. Eng. Mater. Struct.* **2000**, *23*, 667–674. [CrossRef]

10. Akama, M.; Susuki, I. Fatigue crack growth under mixed mode loading in wheel and rail steel. *Tetsu Hagane* **2007**, *93*, 607–613. (In Japanese) [CrossRef]

11. Doquet, V.; Pommier, S. Fatigue crack growth under non-proportional mixed-mode loading in ferritic-pearlitic steel. *Fatigue Fract. Eng. Mater. Struct.* **2004**, *27*, 1051–1060. [CrossRef]

12. Doquet, V.; Bui, Q.H.; Constantinescu, A. Plasticity and asperity-induced fatigue crack closure under mixed-mode loading. *Int. J. Fatigue* **2010**, *32*, 1612–1619. [CrossRef]

13. Dahlin, P.; Olsson, M. The effect of plasticity on incipient mixed-mode fatigue crack growth. *Fatigue Fract. Eng. Mater. Struct.* **2003**, *26*, 577–588. [CrossRef]

14. Yu, X.; Li, L.; Proust, G. Fatigue crack growth of aluminium alloy 7075-T651 under proportional and non-proportional mixed mode I and II loads. *Eng. Fract. Mech.* **2017**, *174*, 155–167. [CrossRef]

15. Bold, P.E. Multiaxial Fatigue Crack Growth in Rail Steel. Ph.D. Thesis, University of Sheffield, Sheffield, UK, 1990.

16. Bower, A.F. The influence of crack face friction and trapped fluid on surface initiated rolling contact fatigue cracks. *ASME J. Tribol.* **1988**, *110*, 704–711. [CrossRef]

17. Kaneta, M.; Matsuda, K.; Murakami, K.; Nishikawa, H. A possible mechanism for rail dark spot defects. *ASME J. Tribol.* **1998**, *120*, 304–309. [CrossRef]

18. Wong, S.L.; Bold, P.E.; Brown, M.W.; Allen, R.J. Two measurement techniques for determining effective stress intensity factors. *Fatigue Fract. Eng. Mater. Struct.* **2000**, *23*, 657–664. [CrossRef]

19. Richard, H.A.; Schramm, B.; Schirmeisen, N.-H. Cracks on Mixed Mode loading—Theories, experiments, simulations. *Int. J. Fatigue* **2014**, *62*, 93–103. [CrossRef]

20. Gao, H.; Alagok, N.; Brown, M.W.; Miller, K.J. Growth of Fatigue Cracks Under Combined Mode I and II Loads. In *Multiexial Fatigue*; ASTM STP 853; Miller, K.J., Brown, M.W., Eds.; ASTM International: West Conshohocken, PA, USA, 1985; pp. 184–202.

21. Barsoum, R.S. On the use of isoparametric finite elements in linear fracture mechanics. *Int. J. Num. Mech. Eng.* **1976**, *10*, 25–37. [CrossRef]

22. Lemaitre, J.; Chaboche, J.-L. *Mechanics of Solid Materials*; Cambridge University Press: Cambridge, UK, 1990.

23. Fujii, Y.; Maeda, K.; Otsuka, A. A new test method for mode II fatigue crack growth in hard materials. *J. Soc. Mater. Sci. Jpn.* **2001**, *50*, 1108–1113. (In Japanese) [CrossRef]

24. Li, C. Vector CTD criterion applied to mixed-mode fatigue crack growth. *Fatigue Fract. Eng. Mater. Struct.* **1989**, *12*, 59–65. [CrossRef]

Article

Fatigue Crack Growth under Non-Proportional Mixed Mode Loading in Rail and Wheel Steel Part 2: Sequential Mode I and Mode III Loading

Makoto Akama [1,*] and Akira Kiuchi [2]

[1] Department of Mechanical Engineering for Transportation, Osaka Sangyo University, Osaka 574-8530, Japan
[2] Formerly Kobelco Research Institute Inc., Hyogo 651-0073, Japan
* Correspondence: akama@tm.osaka-sandai.ac.jp; Tel.: +81-72-875-3001; Fax: +81-72-871-1289

Received: 3 June 2019; Accepted: 15 July 2019; Published: 18 July 2019

Abstract: Rolling contact fatigue cracks in rail and wheel undergo non-proportional mixed mode I/II/III loading. Fatigue tests were performed to determine the coplanar and branch crack growth rates on these materials. Sequential and overlapping mode I and III loading cycles were applied to single cracks in round bar specimens. Experiments in which this is done have been rarely performed. The fracture surface observations and the finite element analysis results suggested that the growth of long (does not branch but grown stably and straight) coplanar cracks was driven mainly by mode III loading. The cracks tended to branch when increasing the material strength and/or the degree of overlap between the mode I and III loading cycles. The equivalent stress intensity factor range that can consider the crack face contact and successfully regressed the crack growth rate data is proposed for the branch crack. Based on the results obtained in this study, the mechanism of long coplanar shear-mode crack growth turned out to be the same regardless of whether the main driving force is in-plane shear or out-of-plane shear.

Keywords: non-proportional mixed mode loading; fractography; mode III stress intensity factor; FEA; rail steel; wheel steel

1. Introduction

The repeated passage of train wheels over the rails induces rolling contact fatigue (RCF) cracks on both the railhead and the wheel tread. Such surface RCF cracks undergo non-proportional mixed mode I/II/III loading cycles [1–4]. For a period, they grow stably at a shallow angle to the surface, according to what is considered coplanar fatigue crack growth (FCG). Once they have reached a certain length, these cracks start to branch. Coplanar cracks are not a great danger to trains as they flake off, causing the train to have a rougher ride at most. Branch cracks, instead, could lead to catastrophic failure if left to grow. However, the continued growth of coplanar cracks can obscure the branch ones from detection through conventional non-destructive methods. Therefore, the accurate prediction of growth rates for both coplanar and branch cracks is essential to prevent rail and wheel failures and develop cost-effective maintenance strategies.

Many studies have been conducted on the coplanar FCG of rail steel under sequentially applied mode I and II loading cycles. Compared to these mixed mode loading tests, FCG experiments under mixed mode I/III loading are hardly performed. However, several tests under proportional loading or with one mode cyclic and other static have been conducted. Ritchie et al. [5] studied the FCG in mode III AISI 4340 steel in torsional loading, measured on circumferentially-notched specimens and the results were compared with the behavior in mode I. FCG rates in mode III were found to be slower than those in mode I, despite that they were still Paris-type law related to mode III stress intensity K_{III} range (ΔK_{III}). They proposed a micromechanical model for mode III FCG, in which the crack advance was considered to occur via a mode II coalescence of cracks, initiated at inclusions ahead of the main crack front. Nayeb-Hashemi et al. [6] found no correlation between the FCG rate and the ΔK_{III} or the plastic strain intensity range ($\Delta\Gamma_{III}$) except when superimposing a static mode I loading to the mode III one on a low-strength AISI 4140 steel. In the latter case, the variability of the FCG rate decreased. Hourlier and Pineau [7] investigated the effects of static mode III on mode I FCG behavior in four materials. They showed that the loading conditions caused two main effects—a considerable reduction in FCG rate and a modification in crack path. They introduced a new criterion based on two main assumptions—fatigue cracking occurred only under the effect of local mode I opening and a fatigue crack grows in a direction where the FCG rate is the maximum. Tarantino et al. [8] applied a novel experimental method to promote mode III coplanar FCG in bearing steel. The method comprised out-of-phase multiaxial fatigue tests, adopting a stress history typical of sub-surface RCF onto specimens containing micronotches. Their analytical model has shown that the typical crack opening values determined by out-of-phase loads can prevent the crack face contact during the RCF loading cycles. Giannella et al. [9] investigated the FCG behavior in cruciform specimens made of Ti6246 by applying a static load along one arm of the specimen and a cyclic load along the other arm. They used an equivalent stress intensity range in the Walker crack growth law that can consider all mode I/II/III loading and determined that a change in the FCG direction occurred depending on the static load levels.

Recently, mixed mode II/III experiments were also performed on a rail steel considering RCF. Bonniot et al. [10] performed the experiments to determine the FCG thresholds and kinetics under loading conditions using asymmetric four-point bending specimens with different angles between the crack front and the shear load. They determined that a coplanar shear-mode FCG occurred with the high loading ranges. The effective stress intensity factors (SIFs) were derived by an inverse method using the relative displacements of the measured crack face. They were found to be 10%–70% lower than the nominal SIF, and a reasonable correlation of the measured FCG rates could be made using the effective SIF.

Akama and Kiuchi [11,12] conducted fatigue tests to determine the coplanar FCG rate on both rail and wheel steel under non-proportional mixed mode I/III loading cycles. They could induce long coplanar cracks under this condition. Moreover, the cracks tended to branch when the degree of overlap (δ) between the mode I and III cycles increased.

As mentioned above, the previous studies about FCG under non-proportional mixed mode loading cycles were apparently focused on the mode I/II loading and coplanar growth in rail steel. Only a detailed study on coplanar and branch FCG under non-proportional mixed mode I/III loading cycles has been performed, by Akama and Kiuchi [11,12], for rail and wheel steel. However, the elucidation of the phenomena inherent to FCG under non-proportional mixed mode I/III loading is not sufficient in their studies. Therefore, in this study, the FCG data obtained from non-proportional mode I/III loading cycles were re-constructed by using novel and reliable equivalent SIF ranges to obtain good correlations. In addition, a finite element analysis (FEA) was performed thoroughly to investigate the crack behavior under these loading cycles.

This paper, which is part 2 of two companion papers, presents FCG under non-proportional mixed mode I/III loading and is organized as follows. Section 2 describes the detailed experimental methods and presents the experimental results. In Section 3, the FEA model for predicting the crack path direction is presented, and the results are provided. Section 4 provides detailed considerations and a discussion by comparing the experimental and FEA results. Finally, the results obtained herein are summarized in Section 5.

2. Experiments

2.1. *Testing Machine and Specimen*

The fatigue tests were performed on a servo-hydraulic testing machine, in which capabilities are a maximum tension–compression force of 200 kN, a maximum fully reversed torque of 1000 Nm under load control, and a frequency of 1 Hz in dry conditions.

Circumferentially notched round bar specimens with a 45° starter notch were used. The detailed geometry is depicted in Figure 1. The notch had a 30° groove, a 4.5 mm depth, and a 0.1 mm root radius and was produced at the specimen center by spark erosion. Precracking was not performed, but the growth data within 0.7 mm from the notch tip were discarded.

Four displacement gauges were placed on the knife edges. These edges were attached across the notch on the specimen surface along the circumferential direction at every 90° to measure the compliance, as shown in Figure 2. The specimen was placed into the machine by adjusting the maximum difference from the average value of the strain obtained from these four clip gauges adjusted to within 5%. Prior to the experiments, specimens containing notch depths of 5, 6, 7, and 8.5 mm were prepared, and the relation between load and output of the displacements from the gauges when each specimen was loaded up to 30 kN was investigated. In each case, a linear relation was obtained with no deviation, which was considered to be caused by slip. Additionally, the accuracy was verified using the potential drop method. A personal computer controlled the testing machine by generating the non-proportional loading cycles, as shown in Figure 3, and recorded the data from the strain gauges. The crack length (or depth) (*a*) was calculated via the compliance technique.

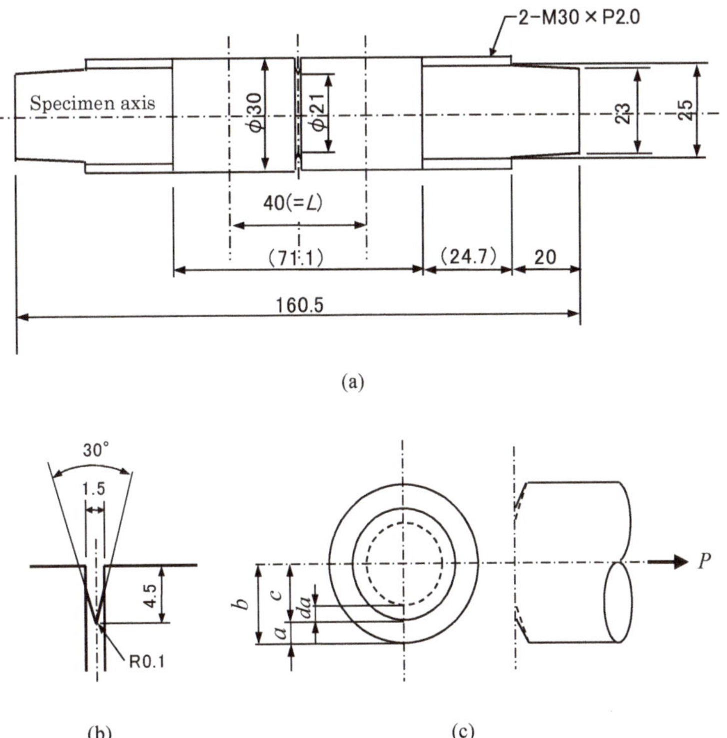

Figure 1. (**a**) Whole configuration of the round bar specimen, (**b**) detail of the notch, and (**c**) parameters of the cracked section (all dimensions in mm).

Figure 2. Set up of the specimen with displacement gauge attached for the measurement of opening displacements.

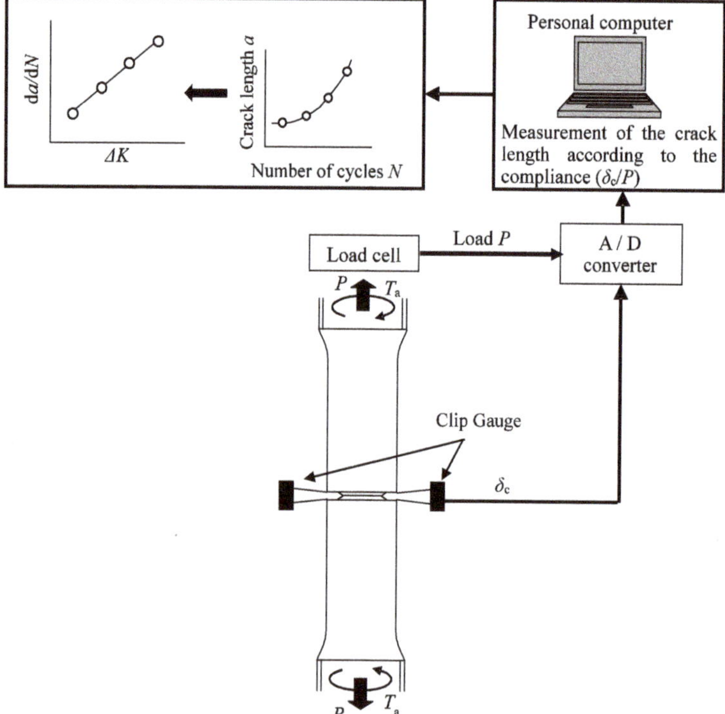

Figure 3. Measurement system for the fatigue crack growth rates.

Two rail steels (RP and RF) and one wheel steel (WT) were used as specimen materials. RP and RF were selected to simulate normal and head hardened rails, respectively. Their chemical compositions and mechanical properties are summarized in Tables 1 and 2. The microstructures of RP and WT were normal pearlite, whereas that of RF was fine pearlite with an average lamellar spacing of about 100 nm. The specimens were directly collected from real rails and wheels.

Table 1. Chemical composition (wt%).

Material	C	Si	Mn	P	S
Rail steel, RP	0.68	0.26	0.93	0.016	0.01
Rail steel, RF	0.79	0.17	0.82	0.019	0.01
Wheel steel, WT	0.65	0.26	0.73	0.016	0.01

Table 2. Mechanical properties.

Material	Ultimate Tensile Strength (MPa)	0.2% Proof Stress (MPa)
Rail steel, RP	934	511
Rail steel, RF	1214	802
Wheel steel, WT	from 981 to 1030	from 618 to 657

2.2. Loading History

The loading history simulated the one experienced by RCF cracks in the presence of a fluid, as obtained by FEA [1,3] and represented in Figure 4, which also shows δ between K_I and K_{III}.

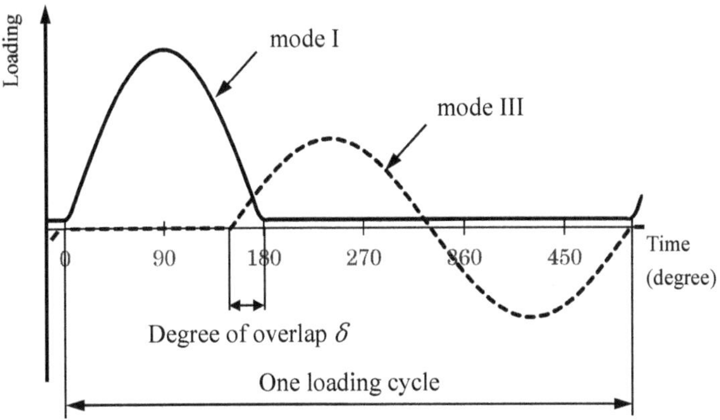

Figure 4. A loading cycle applied under the condition of $\Delta K_{III}/\Delta K_I = 1.0$ and $\delta = 30°$.

2.3. Calculation of the Stress Intensity Factors

K_I and K_{III} were calculated as follows [13]:

$$K_I = f(x)\sigma\sqrt{\pi a} \tag{1}$$

and

$$K_{III} = g(x)\tau\sqrt{\pi a} \tag{2}$$

where σ and τ are the normal and out-of-plane shear stress, respectively, applied to the crack. The functions $f(x)$ and $g(x)$ are expressed as follows:

$$f(x) = \frac{1}{2}\sqrt{x}\left(1 + \frac{1}{2}x + \frac{3}{8}x^2 - 0.363x^3 + 0.731x^4\right) \tag{3}$$

and

$$g(x) = \frac{3}{8}\sqrt{x}\left(1 + \frac{1}{2}x + \frac{3}{8}x^2 + \frac{5}{16}x^3 + \frac{35}{128}x^4 + 0.208x^5\right) \tag{4}$$

where x indicates the *c/b* represented in Figure 1c, *b* is the radius of the specimen, and *c* is the radius of the remaining ligament.

2.4. Experimental Conditions

The experiments were conducted on five RP—three RF and one WT specimen—hereafter referred to as RP1, RP2, RP3, RP4, and RP5; RF1, RF2, and RF3; and WT1. Table 3 reports the $\Delta K_{III}/\Delta K_I$ ratios and δ values for each experiment. In all tests, the crack faces were slightly pulled apart so that the stress ratios of mode I (R_I) and III (R_{III}) loading were 0.05 and −1, respectively. RP1, RP2, RP3, RF1, RF2, and WT1 were designed to provide mainly coplanar FCG rate data, whereas RP4, RP5, and RF3 were conducted to obtain branch FCG rate data. RP1 was carried with $\delta = 90°$ at the first step. Once the crack length reached about 7 mm, it was interrupted and continued by changing δ to 30°. In a similar way, RP3, RF1, RF2, and WT1 were performed by changing δ. After the tests, the specimens were broken by applying a large tensile force to observe the fracture surfaces.

Table 3. Testing conditions for mixed mode I/III loading tests.

Exp. No.	$\Delta K_{III}/\Delta K_I$	Δ (degree)
RP1	1.0	90 → 30
RP2	1.5	90
RP3	1.5	30 → 120
RP4	1.0	180
RP5	1.5	180
RF1	1.0	90 → 30
RF2	1.5	60 → 30
RF3	1.0	180
WT1	1.0	90 → 60

2.5. Experimental Results

Figure 5 schematically represents the main coplanar crack plane and the angles between this plane and the branch crack plane when branching occurred. In this figure, θ and φ are the twisting and deflection angles, respectively.

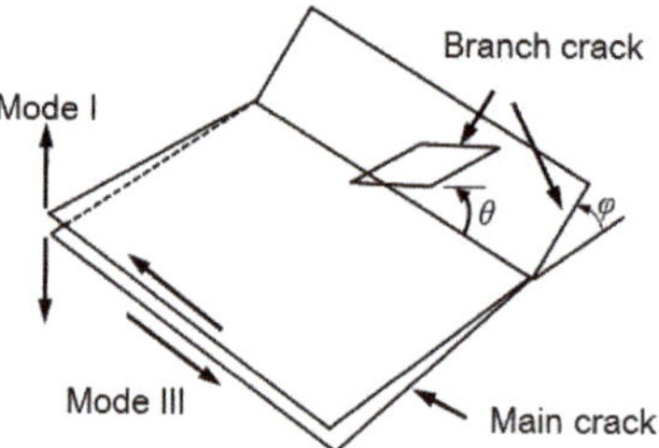

Figure 5. Branch crack at a main crack and branch angle definition.

The coplanar cracks grew almost straight (i.e., $\theta = \varphi = 0°$) in RP1 and RP2 but branched in RP3 at about $\theta = 45°$ and $\varphi = 0°$, when δ was 120°. In RP4 and RP5, the branching occurred and resulted in factory-roof fractures (i.e., associated with ridges and valleys). In RF1 and RF2, little factory-roof fractures were observed on the crack faces when δ was 90° and 60°, respectively, but coplanar cracks grew in both experiments when it was reduced to 30°. The crack branched in RF3. In WT1, the fracture surface was flat, the crack grew coplanarly, and no branching was observed. Figure 6 shows the macroscopic fracture surfaces obtained in RP2, RP5, RF2, and WT1.

2.5.1. Coplanar Crack Growth Rate

When mode III and mode I are mixed, the equivalent SIF range at $\varphi = 0°$ in Figure 5 can be expressed as [14]

$$\Delta K_s = 0.5\left\{\Delta K_I^2 (1 - 2v)^2 + 4\Delta K_{III}^2\right\}^{0.5} \tag{5}$$

where v is the Poisson ratio. When $v = 0.3$, it becomes

$$\Delta K_s = 0.5\left\{0.16\Delta K_I^2 + 4\Delta K_{III}^2\right\}^{0.5} \tag{6}$$

Figure 6. Macroscopic fracture surface of (**a**) RP2 ($\Delta K_{III}/\Delta K_I = 1.5$, $\delta = 90°$), (**b**) RP5 ($\Delta K_{III}/\Delta K_I = 1.5$, $\delta = 180°$), (**c**) RF2 ($\Delta K_{III}/\Delta K_I = 1.5$, $\delta = 60°\to30°$) and (**d**) WT1 ($\Delta K_{III}/\Delta K_I = 1.0$, $\delta = 90°\to60°$).

Figure 7 shows the growth rate data obtained for RP, consisting of the RP1, RP2, and RP3 results, with respect to ΔK_s. The data are observed to be divided into two groups by δ. When the Paris-type law was applied to the data for the same δ, the following equations were obtained:

$$\frac{da}{dN} = C(\Delta K_s)^m \tag{7}$$

where N is the number of cycles. For $\delta = 90°$, $C = 2.52 \times 10^{-11}$ and $m = 2.93$ and for $\delta = 30°$, $C = 1.15 \times 10^{-10}$ and $m = 2.26$. The R-squared value, i.e., the coefficient of determination (R^2), is 0.973 and 0.917. In each case, a good correlation could be obtained even when the $\Delta K_{III}/\Delta K_I$ differed.

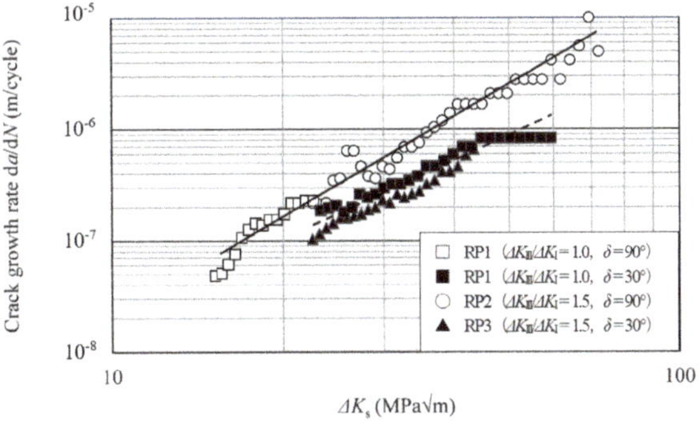

Figure 7. Coplanar crack growth rates against ΔK_s for RP.

Next, the growth rate data obtained for RF, comprising of the RF1 and RF2 results, were correlated with respect to ΔK_s, as shown in Figure 8. In this case, the correlation was poor. The growth law was as follows:

$$\mathrm{d}a/\mathrm{d}N = 7.40 \times 10^{-10}(\Delta K_s)^{1.62} \ (R^2 = 0.638) \tag{8}$$

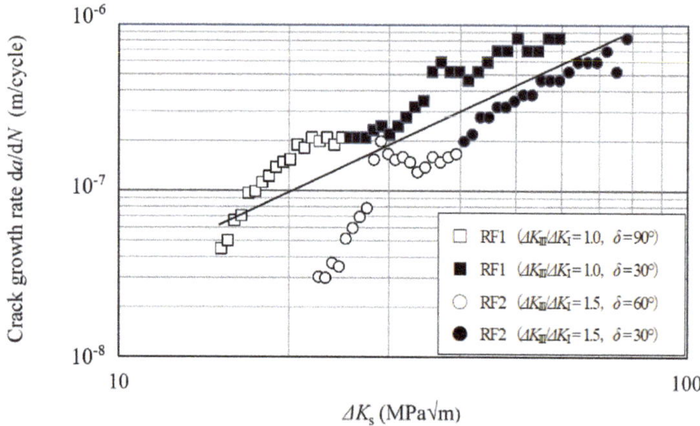

Figure 8. Coplanar crack growth rates against ΔK_s for RF.

Although not shown in the figure, when δ was constant, the growth rates increased with $\Delta K_{\mathrm{III}}/\Delta K_{\mathrm{I}}$. When $\Delta K_{\mathrm{III}}/\Delta K_{\mathrm{I}}$ was constant, the growth became faster owing to the increased δ.

Figure 9 shows the FCG rate data as a function of ΔK_s when the $\Delta K_{\mathrm{III}}/\Delta K_{\mathrm{I}}$ was 1.0, for all the specimens. WT exhibited the highest growth rate, followed by RP and RF.

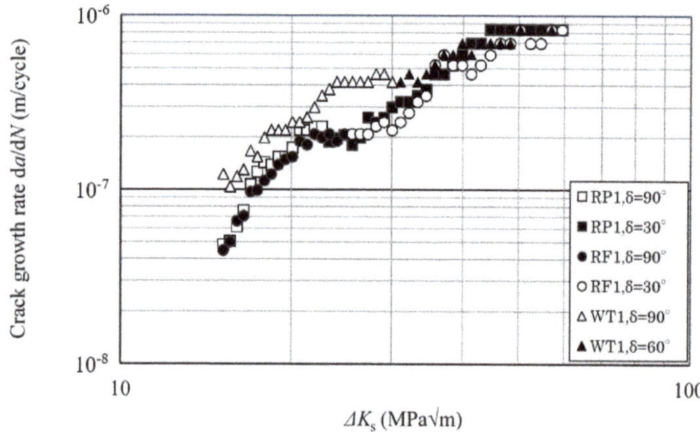

Figure 9. Comparison of crack growth rates against ΔK_s for RP, RF, and WT ($\Delta K_{\mathrm{III}}/\Delta K_{\mathrm{I}} = 1.0$).

2.5.2. Branch Crack Growth Rate

In the mixed mode I/III loading, the cracks grown may branch to form torsional facets. In such a case, the cracks form perpendicularly to the maximum principal stress and the equivalent SIF range has been proposed as follows [15]:

$$\Delta K_t = 0.8\Delta K_{\mathrm{I}} + 0.5\left\{0.16\Delta K_{\mathrm{I}}^2 + 4\Delta K_{\mathrm{III}}^2\right\}^{0.5} \tag{9}$$

under the condition of $v = 0.3$. The combined growth data obtained from RP4 and RP5 as a function of ΔK_t are plotted in Figure 10. When $\Delta K_{III}/\Delta K_I$ was large (i.e., in the RP5 case), the growth rates went down. This happened because Equation (9) does not consider the attenuation of ΔK_{III} due to the crack face contact. This attenuation was considered in the previously performed mixed mode I/III test with $R_I = R_{III} = 0.05$ and the equivalent SIF range (ΔK_{eq}) that was weighted to ΔK_{III}, as in the following equation [16], was proposed:

$$\Delta K_{eq} = \left\{ \Delta K_I^2 + 0.2\Delta K_{III}^2 \right\}^{0.5} \tag{10}$$

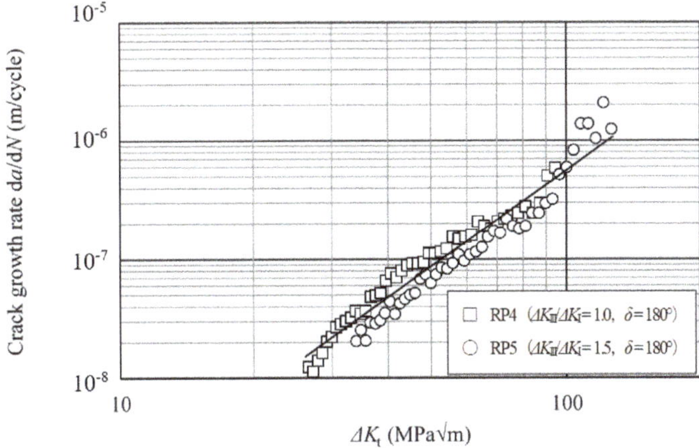

Figure 10. Branch crack growth rates against ΔK_t for RP.

When the growth rates were plotted against ΔK_{eq}, a good correlation was obtained even when $\Delta K_{III}/\Delta K_I$ and δ were extensively changed. Therefore, also in this study, the growth rates for RP and RF were plotted against ΔK_{eq}, and the results for RP are shown in Figure 11, revealing a fairly good correlation. The growth rate in RP was expressed via the Paris-type law as follows:

$$da/dN = 1.84 \times 10^{-12}(\Delta K_{eq})^{3.29} \ (R^2 = 0.960) \tag{11}$$

Although not shown, the growth law for RF was

$$da/dN = 4.66 \times 10^{-13}(\Delta K_{eq})^{3.71} \ (R^2 = 0.987) \tag{12}$$

As mentioned previously, coplanar growth was considered to occur in RF1 and RF2 when δ was 30°, and the growth rates were plotted with ΔK_s. However, the correlation was poor in these cases. This was probably due to the effect of small factory-roof fractures that formed on the fracture surfaces under the initial conditions. Although the coplanar growth occurred when δ was 30° for both specimens, an effect of the contact at the factory-roof fracture surfaces on the growth rates was observed throughout the experiments. Therefore, all data were correlated with ΔK_{eq}, and the results are shown in Figure 12. A better correlation was obtained compared with the case of plotting with ΔK_s. The growth law was as follows:

$$da/dN = 4.51 \times 10^{-11}(\Delta K_{eq})^{2.60} \ (R^2 = 0.823) \tag{13}$$

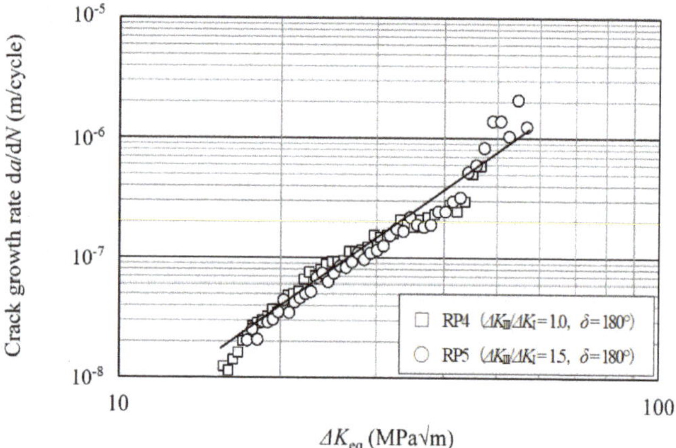

Figure 11. Branch crack growth rates against ΔK_{eq} for RP.

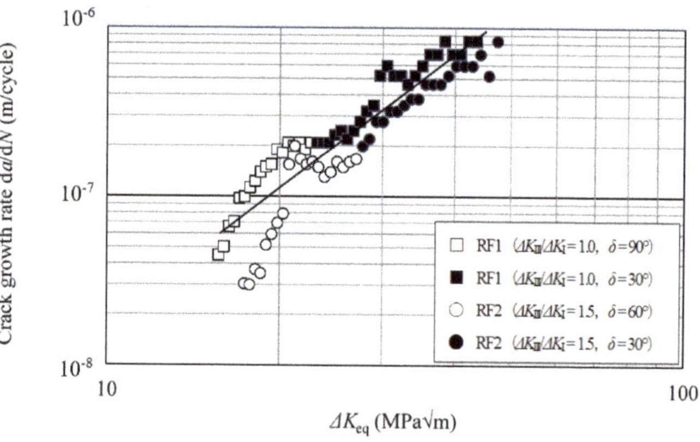

Figure 12. Branch crack growth rate against ΔK_{eq} for RF.

2.5.3. Fractography

After the experiments, scanning electric microscope (SEM) observations were performed to unravel the FCG mechanism from the fracture surfaces of the specimens. The SEM images of the fracture surfaces of the RP1, RP2, and RP3 specimens, at various growth times, are shown in Figures 13–15. Fujii et al. [17] performed fatigue tests under mode II loading, with high growth rates and short crack lengths, and observed severe rub marks and many wear particles and oxides on the fracture surfaces. In RP1 and RP2, however, there was no evidence of contact between the crack surfaces at all growth periods, and no oxide debris generated from the surface appeared. At high magnifications, microcleavage and rippling could be observed, implying that the attenuation due to surface friction was very limited. As in previous non-proportional mixed mode I/II loading experiments [18,19], clear striation patterns were seldom seen. Because there was no contact between the crack faces, the striation patterns were not worn out by the mating face.

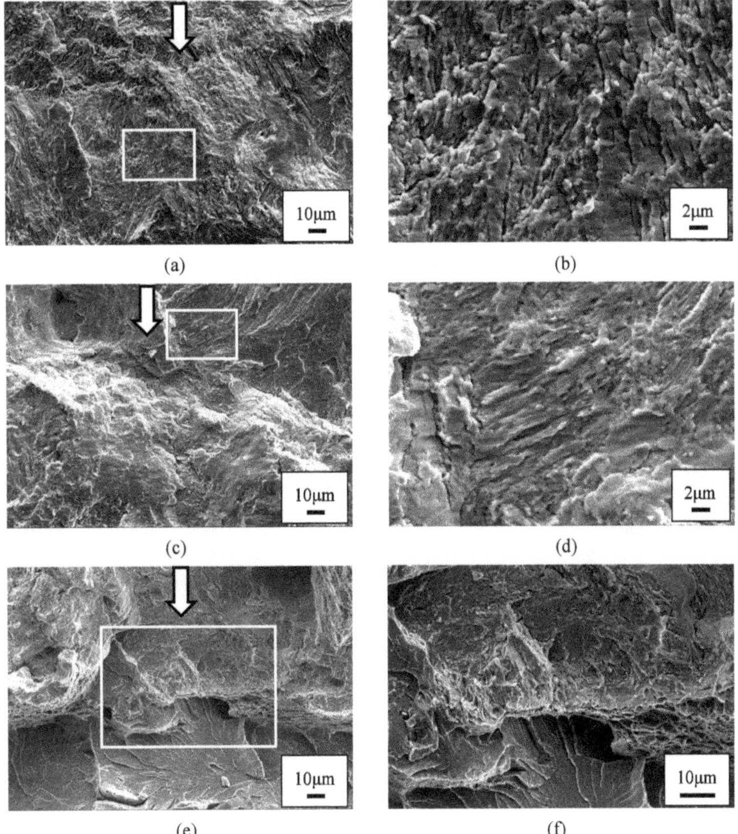

Figure 13. Fracture surface of RP1 observed via scanning electric microscope (SEM) (arrows indicate the direction of the crack growth). (**a**) 2 mm from the notch tip ($\delta = 90°$), (**b**) an enlarged view indicated by the frame in (**a**), (**c**) 4.8 mm from the notch tip ($\delta = 30°$), (**d**) frame in (**c**), (**e**) 10.5 mm from the notch tip ($\delta = 30°$), and (**f**) frame in (**e**).

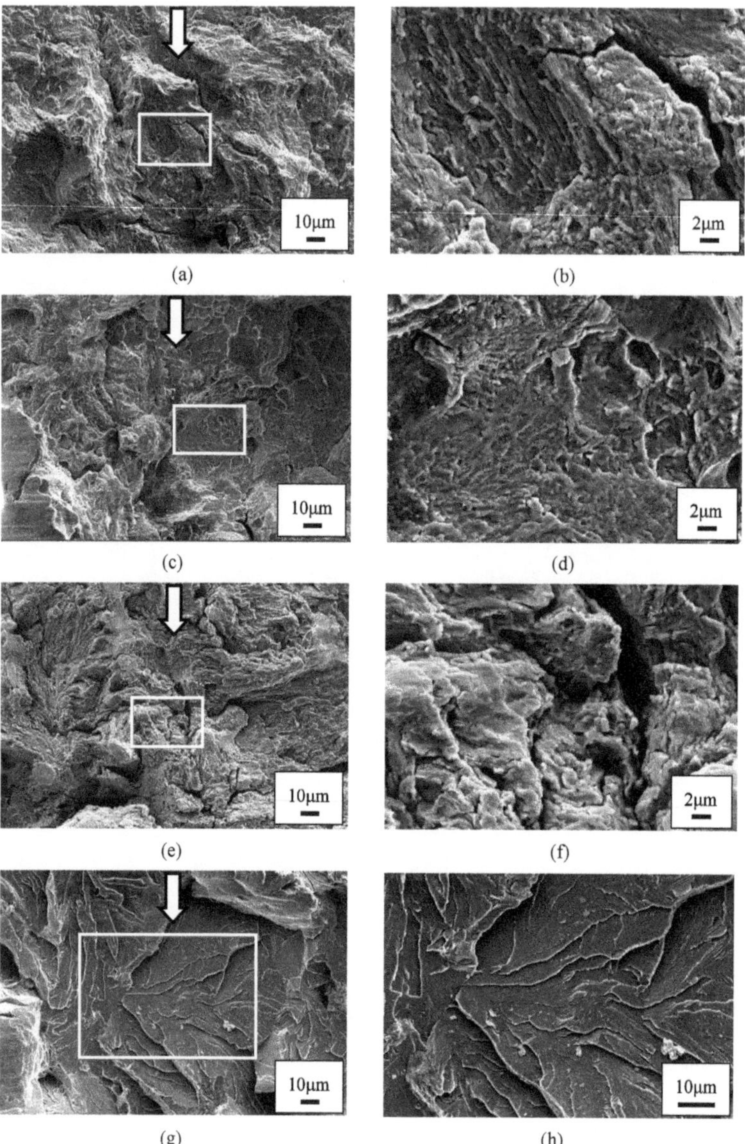

Figure 14. Fracture surface of RP2 observed via SEM (arrows indicate the direction of the crack growth). (**a**) 2 mm from the notch tip, (**b**) an enlarged view indicated by the frame in (**a**), (**c**) 4 mm from the notch tip, (**d**) frame in (**c**), (**e**) 6.8 mm from the notch tip, (**f**) frame in (**e**), (**g**) 8 mm from the notch tip, and (**h**) frame in (**g**).

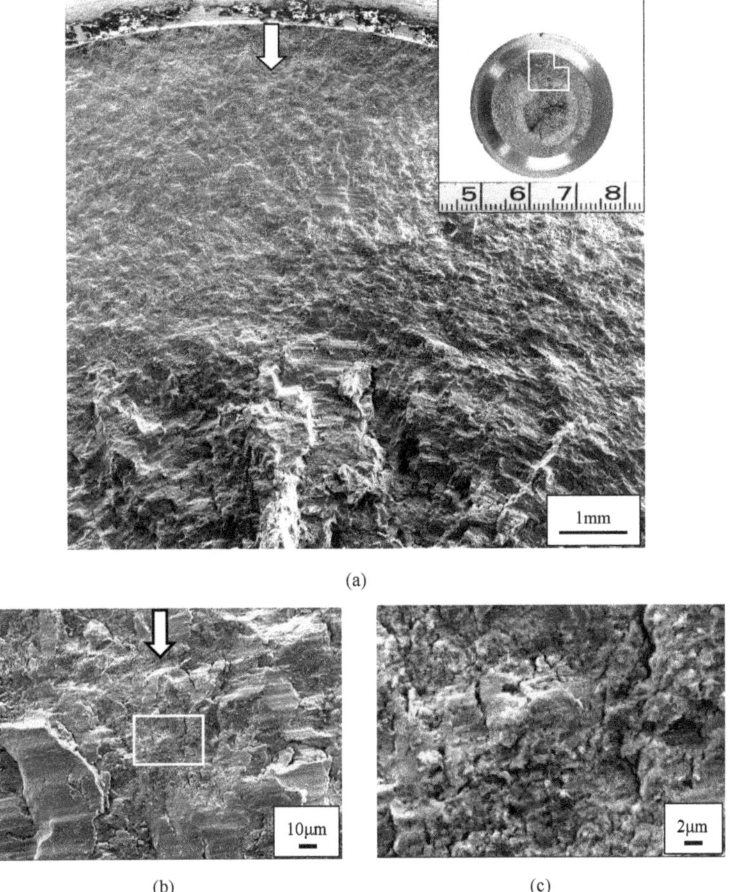

(a)

(b)　　　　　　　　　　　　　(c)

Figure 15. Fracture surface of RP3 observed using a SEM (arrows indicate the direction of the crack growth). (**a**) Macroscopic appearance of the fracture surface, (**b**) 3 mm from the notch tip ($\delta = 30°$), and (**c**) an enlarged view indicated by a frame in (**b**).

3. Finite Element Analysis

3.1. Procedure

A series of elasto-plastic FEA was performed using the commercial FEA code MARC to elucidate the FCG characteristics observed in the mixed mode I/III loading tests and predict the FCG direction, which is θ and φ in Figure 5. Therefore, a stationary crack was considered. The three-dimensional (3D) finite element mesh of the round bar specimen and the boundary conditions applied are shown in Figure 16. The plasticity was considered highly localized near the notch under the loading, while other specimen regions remained elastic. Therefore, the area close to the symmetry section for the axial direction was enough for the mesh. The mesh density increased at the crack tip where, for the radial direction, the size of an 8-node brick element was 10 μm, which was considered well included in the plastic zone ahead. The total numbers of elements and nodes were 14,001 and 15,522, respectively. This mesh and the loading procedure confirmed that the nominal stress value in the ligament ahead of the crack was accurate.

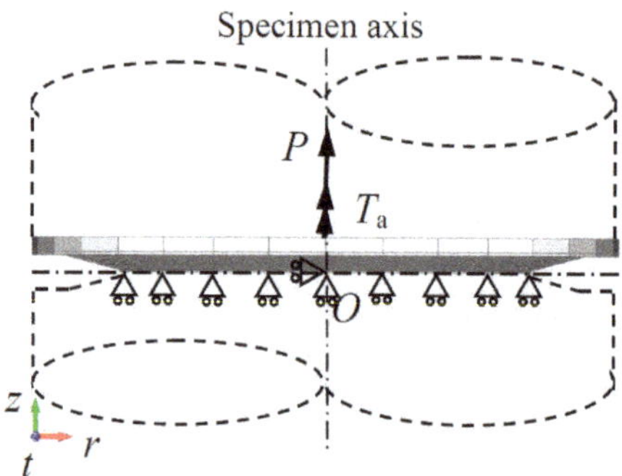

Figure 16. Mesh used for the analyses, and applied loads and boundary conditions.

At the center of the upper surface of the mesh, a pilot node was connected to the remaining nodes of that surface through a multipoint constraint; force (P) and torque (T_a) were applied to the node; their ranges ΔP and ΔT_a were equal to 55 kN ($R_I = 0.05$) and 525 Nm ($R_{III} = -1$), respectively. If ΔP and ΔT_a were equal to 55 kN and 525 Nm, respectively, $\Delta K_{III}/\Delta K_I$ became 1.5. The fixed boundary conditions applied on the lower surface of the mesh were of the same as the symmetry condition for the axial direction. The a value was set at 4.6 mm for the cases of $\delta = 30°$, $90°$, and $180°$ and at 7 mm for $\delta = 120°$, representing the actual RP3 test.

In this study, the model combining nonlinear kinematic hardening rule with the isotropic hardening rule developed by Chaboche and Lemaitre [20] (C & L model) was employed.

$$^{ti+\Delta ti}\sigma_y = {}^0\sigma_y + Q\left\{1 - \exp\left(-B^{ti+\Delta ti}\bar{e}^p\right)\right\} \tag{14}$$

and

$$d\alpha = \frac{2}{3}hde^p - \zeta\alpha d\bar{e}^p \tag{15}$$

where $^{ti+\Delta ti}\sigma_y$ is the updated yield stress at time $t_i + \Delta t_i$, $^0\sigma_y$ is the initial yield stress, Q, B, h, and ζ are material constants, $^{ti+\Delta ti}\bar{e}^p$ is the accumulated effective plastic strain at $t_i + \Delta t_i$, α is the shift of the yield surface center, e^p is the plastic strain, and d implies increment. The material constants for RP and RF are summarized in Table 4 along with Young's modulus and Poisson ratio. The FEA was performed on these two rail steels to clarify the material effect on the FCG rate.

Table 4. Material properties used in finite element analysis (FEA).

Material	E (MPa)	ν	$^0\sigma_y$ (MPa)	Q	b	h (MPa)	ζ
RP	183,008	0.3	508	−208	24.2	85,248	193
RF	182,778	0.3	684	−264	1.27	88,615	185

3.2. Analytical Results

The contour plots of total displacement and *tz*-stress on the shapes deformed under the positions of the loading cycle for the RP case are shown in Figure 17. The *rtz* represents global cylindrical coordinate system whose origin coincides with specimen center O. The following evaluations were performed during the 160th loading cycle because the stress states have converged at that cycle.

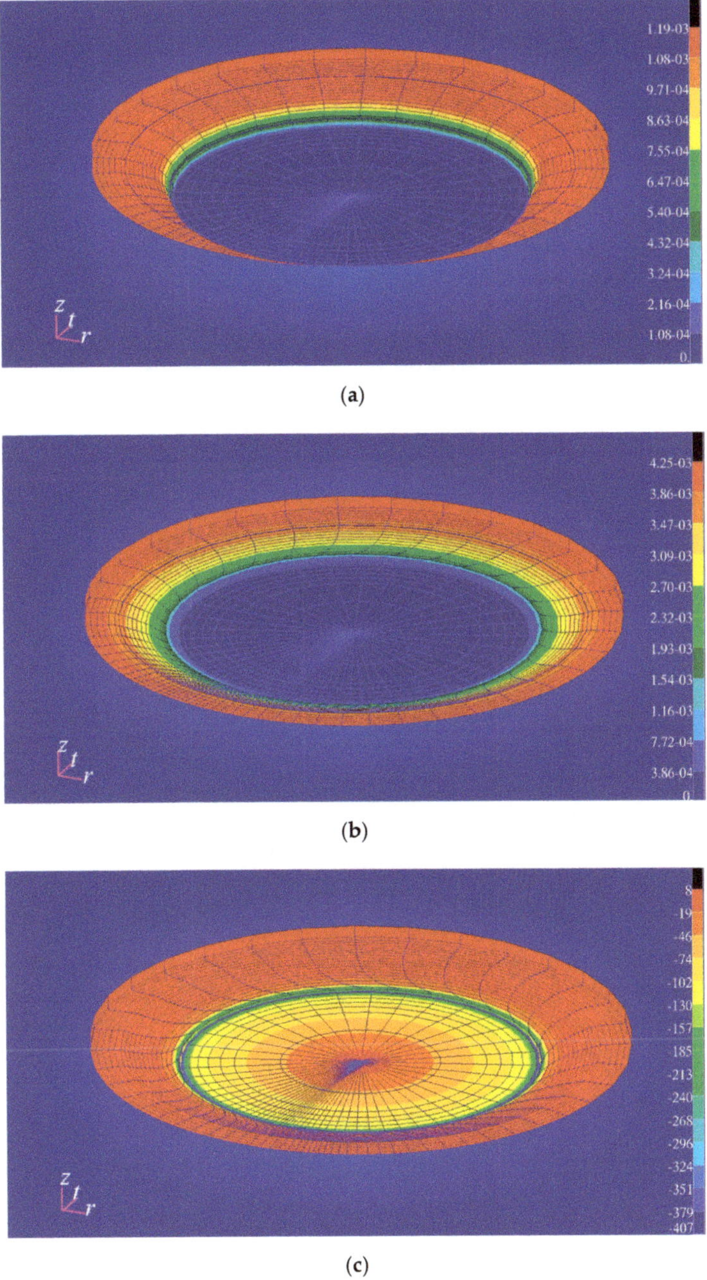

Figure 17. The contour plots of total displacement and *tz*-stress on the shapes deformed under the positions of the loading cycle for the RP case ($\Delta K_{III}/\Delta K_I = 1.5$, $\delta = 90°$), (**a**) maximum deformation by mode I loading, (**b**) and (**c**) maximum deformations by mode III loading. (**a**) and (**b**) depict total displacements and (**c**) depicts *tz*-stress. Notably, the contour levels for displacement in mm and for stress in MPa. Deformation amplifications are 1000×.

3.2.1. Planes of Maximum Normal and Shear Stress Ranges

First, the maximum tangential stress range was investigated to elucidate the effect of δ on crack branching direction [11]. The analysis was suggested to be based on an elasto-plastic stress field [21]. The normal stress range ($\Delta\sigma$) and the shear stress range ($\Delta\tau$) level on every plane were investigated. Such a plane was expressed by θ, φ, and ψ that are the rotation angles around the axes of the local cylindrical coordinate system R, T, and Z whose origin coincides with the center of the crack tip element (see Figure 18). Underlying assumption is that if the cracks grow by a tensile mode, the growth direction should be determined by the $\Delta\sigma$, whereas when the cracks grow by a shear mode, the direction is determined by the $\Delta\tau$ near the crack tip. Considering that $\Delta\tau_{ZR}$ was very small under the applied loading cycles, $\Delta\tau$ was represented by a range of $\Delta\tau_{ZT}$ values.

The variations of $\Delta\sigma$ and $\Delta\tau$ due to θ for the different values of δ for the RP case were indicated on the $\varphi = \psi = 0$ plane in Figure 19. $\Delta\sigma$ and $\Delta\tau$ at each θ were divided by the maximum of $\Delta\tau$ ($\Delta\tau_{max}$) because the absolute values depend on the distance from the crack tip and are not important. When δ increases, the relative $\Delta\sigma$ value also increases and the maximum of $\Delta\sigma$ ($\Delta\sigma_{max}$) plane turns toward the branch direction, whereas the $\Delta\tau_{max}$ plane remains on the coplanar plane. In the calculation results, both $\Delta\sigma_{max}$ and $\Delta\tau_{max}$ planes were slightly oriented toward the φ direction within 7° for all cases.

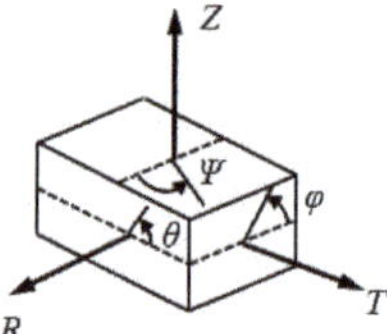

Figure 18. Definition of local cylindrical coordinate system *RTZ*. Notably, θ and φ coincide with the branch angles defined earlier in Figure 5.

3.2.2. Planes of Maximum Principal Stress

The criterion for multiaxial FCG proposed by Schöllmann et al. [22] is based on the maximum principal stress (MPS)—the crack will grow radially from the crack front in the direction that is perpendicular to the MPS, σ_1, on a virtual cylindrical surface around the crack front (see Figure 5). Therefore, herein, the maximum values of σ_1 ($\sigma_{1\,max}$) and directions perpendicular to $\sigma_{1\,max}$ during one loading cycle were investigated for each loading condition.

The $\sigma_{1\,max}$, middle (σ_2), and minimum (σ_3) principal stresses for the different values of δ for the RP case are shown in Figure 20, and in Table 5, the planes of $\sigma_{1\,max}$ are indicated for this case. As shown, as δ increases, the $\sigma_{1\,max}$ plane turns toward the branch direction.

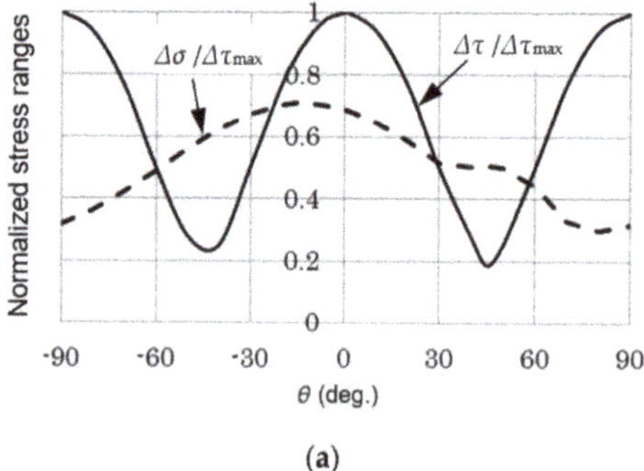

Figure 19. *Cont.*

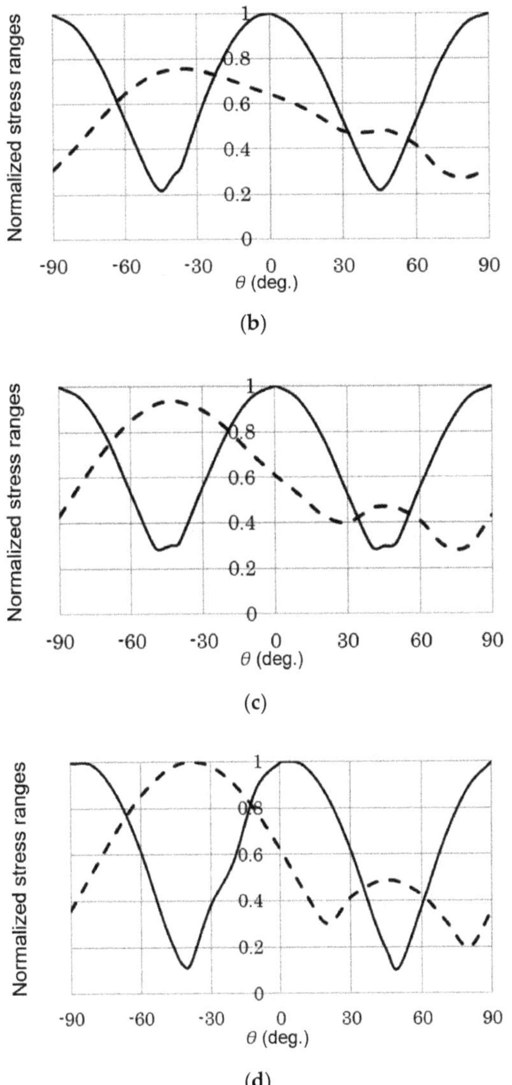

Figure 19. Shear stress range and normal stress range at each angle θ for the case of RP steel ($\Delta K_{\mathrm{III}}/\Delta K_{\mathrm{I}} = 1.5$, $R_{\mathrm{I}} = 0.05$, $R_{\mathrm{III}} = -1$). (**a**) $\delta = 30°$, (**b**) $\delta = 90°$, (**c**) $\delta = 120°$, and (**d**) $\delta = 180°$.

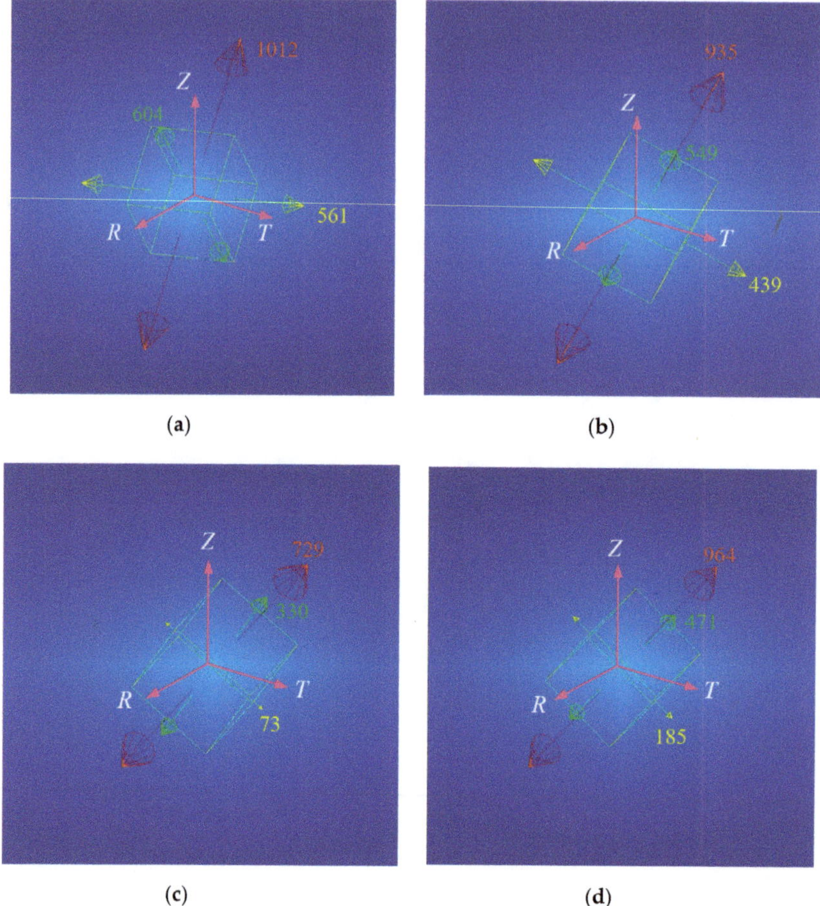

Figure 20. Display of principal stress tensors when σ_1 was the maximum for the case of RP steel ($\Delta K_{III}/\Delta K_I = 1.5$, $R_I = 0.05$, $R_{III} = -1$). (**a**) $\delta = 30°$, (**b**) $\delta = 90°$, (**c**) $\delta = 120°$, and (**d**) $\delta = 180°$. Red arrow indicates $\sigma_{1\,max}$, and green and yellow arrows indicate σ_2 and σ_3, respectively. Their lengths represent the relative magnitudes and the figures are in MPa.

Table 5. Planes of $\sigma_{1\,max}$ for the case of RP steel ($\Delta K_{III}/\Delta K_I = 1.5$, $R_I = 0.05$, $R_{III} = -1$).

δ(deg.)	30	90	120	180
θ(deg.)	−17	−27	−42	−41
φ(deg.)	−4	−3	9	4

3.2.3. Crack Tip Opening Displacement

The FEA results were used to clarify the crack tip opening displacement (CTOD) of RP and RF during the 160th loading. Figure 21 shows the variations of CTOD/2 at several positions from the crack tip (D) for both steels. The values are smaller for RF; in particular, when the tensile load decreased, the time in degrees was greater than 180°, half of the CTODs became smaller than 4×10^{-5} mm in a wide range (50 μm $\leq D \leq$ 100 μm).

4. Discussion

In the case of RP and WT, a coplanar growth was obtained when δ was smaller than or equal to 90° regardless of the $\Delta K_{III}/\Delta K_I$ value and the growth rate was well correlated with ΔK_s, as shown in Figure 9. According to Equation (6), ΔK_I was equal to only 4% of ΔK_{III}, suggesting a strong dependence of the FCG rate on the mode III loading. Conversely, in the RF case, some factory-roof fractures were observed at $\delta = 60°$ and 90°, and no good correlation with ΔK_s was obtained.

The FEA results shown in Figure 21 indicated that the CTOD variation in one loading cycle was smaller for RF than for RP under the same loading condition. In non-proportional mixed mode loading cycles, fatigue cracks follow the direction of the $\Delta\sigma_{max}$ or $\Delta\tau_{max}$ plane depending on if these growth rates on these planes is faster [23]. When the CTOD is small, the actual crack faces have irregularities and, hence, are likely to make contact. When the crack faces make contact, the friction can make the coplanar FCG rate smaller relative to the branch FCG rate. Therefore, factory-roof fractures, which are traces of branch FCG, should have appeared.

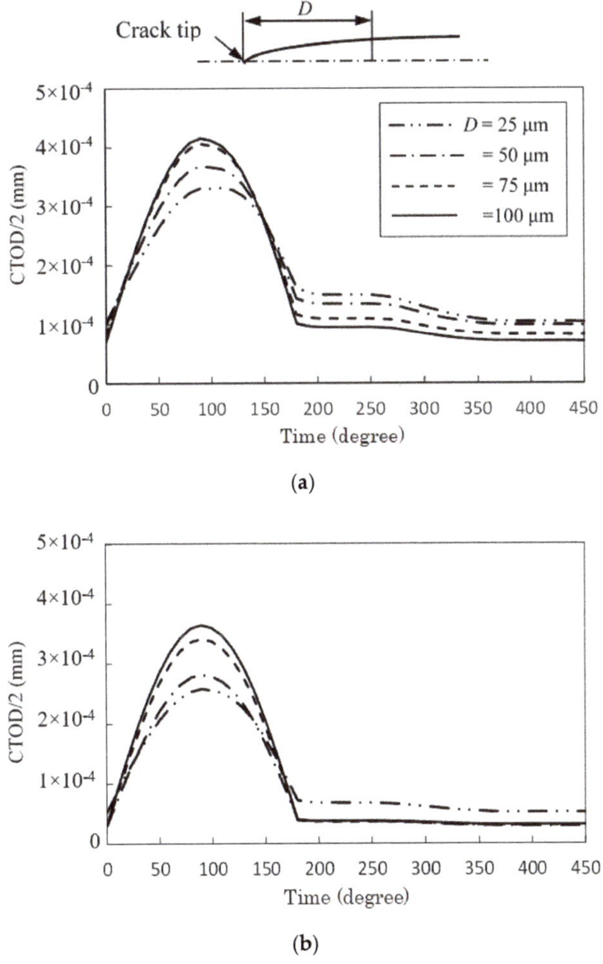

Figure 21. Variations of half the crack tip opening displacements behind the crack tip at 160th cycle of (a) RP and (b) RF. ($\Delta K_{III}/\Delta K_I = 1.5$, $\delta = 90°$).

When δ increased to 120°, branch cracks also grew even in RP. According to the FEA results shown in Figure 19, increasing δ increased $\Delta\sigma_{max}$ with respect to $\Delta\tau_{max}$ and turned it toward the branch direction. Conversely, $\Delta\tau_{max}$ remained on the coplanar direction. The coplanar FCG rates under the loading cycles adopted herein are considered to have a strong dependence on K_{III}. Therefore, it should also have a relatively strong dependence on $\Delta\tau_{max}$, while the branch FCG rates are generally considered to have a relation with $\Delta\sigma_{max}$. These facts conclude that cracks tend to branch when δ increases.

We determined that when δ increases, the $\sigma_{1\,max}$ plane turns toward the branch direction, as indicated in Table 5. However, it may be unsuitable to use the MPS criterion to predict the FCG direction under the loading conditions adopted herein. This criterion assumes that the crack grows in mode I. When the coplanar FCG planes were observed using SEM, no clear striation, which is evidence of FCG caused by mode I loading, was observed.

The SEM observations provided no evidence to support crack face contact in case of RP at $\Delta K_{III}/\Delta K_I = 1.5$ and $\delta = 90°$. This was confirmed by the FEA results, in which the crack was always widely open (see Figure 21a). Rubbing has been suggested as the reason for the FCG termination. Under the testing conditions in the present study, the crack faces were opened during the experiments. If the crack face contacts with its mating face and the generated friction attenuates the mode III loading, the crack is considered to be arrested. If there is no contact, a long shear mode crack becomes possible under this loading condition.

Although the branch FCG rates were plotted against ΔK_t, a good correlation was not obtained because ΔK_t does not consider the ΔK_{III} attenuation due to the crack face contact. Therefore, ΔK_{eq}, which includes this attenuation, was proposed, providing a FCG law with a fairly good correlation. Even for the RF steel, where some factory-roof fractures arose on the fracture surface, a better correlation was obtained when using ΔK_{eq} (see Figure 12). In these cases, the occurrence of some crack face contacts was also considered.

Murakami et al. [24] studied the fatigue crack behavior of S45C steel under pure mode II and mode III loading. Because the fractographic observations after each experiment revealed strong similarities, they concluded that the mechanisms of mode II and III shear FCG are essentially the same. When comparing the fracture surfaces obtained in this study under mixed mode I/III loading with those from mixed mode I/II loading [25], it is clear that they were very similar. In particular, no clear striation patterns were found near the crack tip region in both the cases. Moreover, the coplanar cracks branched when δ was increased to 120° and the loading ratio was 1.5, regardless of $\Delta K_{II}/\Delta K_I$ or $\Delta K_{III}/\Delta K_I$, for both loading cases. WT exhibited the fastest coplanar FCG rates and, among the rail steels, the rates for RP were higher than those for RF when plotted against their appropriate equivalent SIF ranges, for both loading cases. Furthermore, Akama and Kiuchi [11] reported that FCG rates in RP under these two loading conditions (mixed modes I/II and I/III) were almost equal when plotted against the SIF ranges considered to be the main driving forces, ΔK_{II} and ΔK_{III} for the mixed mode I/II and I/III loading conditions, respectively. Therefore, the mechanism of shear-mode FCG under non-proportional mixed mode loadings that were subject to the RCF cracks can be considered to be the same even if the main crack driving force is in-plane shear or out-of-plane shear.

5. Conclusions

To determine the coplanar and branch FCG rates of normal rail, head hardened rail, and wheel steel, fatigue tests were conducted under non-proportional mixed mode I/III loading cycles that simulated the RCF conditions. SEM observations and FEA were also performed to investigate the FCG behavior. The results can be summarized as follows.

1. In RP and WT, a coplanar growth was obtained when δ was smaller than or equal to 90°. The growth rates were relatively well correlated when plotted against ΔK_s defined by Equation (6). The highest coplanar FCG rate was observed in WT, followed by RP and then RF.

2. In the RP case, the branch FCG occurred at $\delta = 120°$. The growth rates were plotted against ΔK_{eq} defined by Equation (10), which considers the ΔK_{III} attenuation due to the crack face contact, giving a good correlation.

3. Based on the fracture surface observations by SEM and the FEA results, the growth of long coplanar cracks was assumed to be driven mainly by mode III loading.

4. The FEA results showed that RF, which is a high-tensile steel, had smaller CTODs during the loading cycles compared with RP. Therefore, contact was likely to occur between the crack faces owing to the surface irregularities, causing crack branching in RF even under the same conditions.

5. When δ increased, $\Delta\sigma_{max}$ with respect to $\Delta\tau_{max}$ also increased, and the $\Delta\sigma_{max}$ plane turned toward the branch direction. Therefore, it can be concluded that the cracks tend to branch when δ increases.

6. The comparison of the fracture surfaces, branching conditions, and coplanar FCG rates data under mixed mode I/III loading and those under mixed mode I/II loading [25] indicated that the coplanar crack growth mechanisms in these two loading cases were similar regardless of whether the main driving force was in-plane or out-of-plane shear.

Author Contributions: Conceptualization, M.A.; Methodology, M.A. and A.K.; Software, M.A.; Validation, M.A. and A.K.; Formal Analysis, M.A.; Investigation, M.A. and A.K.; Writing-Original Draft Preparation, M.A.; Writing-Review & Editing, M.A.; Supervision, M.A.

Funding: This research was performed as part of the research and development programme for the future of railways, entitled 'Creation of rail damage/ballast track deterioration models and evaluation of maintenance work saving technologies', and was funded by the Railway Technical Research Institute.

Conflicts of Interest: The founding sponsors had no role in the design of the study; in the collection, analyses, or interpretation of data; in the writing of the manuscript, and in the decision to publish the results.

References

1. Bogdański, S.; Olzak, M.; Stupnicki, J. Numerical modeling of a 3D rail RCF 'squat'-type crack under operating load. *Fatigue Fract. Eng. Mater. Struct.* **1998**, *21*, 923–935. [CrossRef]

2. Akama, M.; Mori, T. Boundary element analysis of surface initiated rolling contact fatigue cracks in wheel/rail contact systems. *Wear* **2002**, *253*, 35–41. [CrossRef]

3. Bogdański, S.; Lewicki, P. 3D model of liquid entrapment mechanism for rolling contact fatigue cracks in rails. *Wear* **2008**, *265*, 1356–1362. [CrossRef]

4. Akama, M.; Nagashima, T. Some comments on stress intensity factor calculation using different mechanisms and procedures for rolling contact fatigue cracks. *Proc. Inst. Mech. Eng. Part F J. Rail Rapid Transit* **2009**, *223*, 209–221. [CrossRef]

5. Ritchie, R.O.; McClintock, F.A.; Nayeb-Hashemi, H.; Ritter, M.A. Mode III fatigue crack propagation in low alloy steel. *Metall. Trans. A Phys. Metal. Mater. Sci.* **1982**, *13*, 101–110. [CrossRef]

6. Nayeb-Hashemi, H.; McClintock, F.A.; Ritchie, R.O. Effects of friction and high torque on fatigue crack propagation in mode III. *Metall. Trans. A Phys. Metal. Mater. Sci.* **1982**, *13*, 2197–2204. [CrossRef]

7. Houlier, F.; Pineau, A. Propagation of fatigue cracks under polymodal loading. *Fatigue Fract. Eng. Mater. Struct.* **1982**, *5*, 287–302. [CrossRef]

8. Tarantino, M.G.; Beretta, S.; Foletti, S.; Lai, J. A comparison of mode III threshold under simple shear and RCF conditions. *Eng. Fract. Mech.* **2011**, *78*, 1742–1755. [CrossRef]

9. Giannella, V.; Dhondt, G.; Kontermann, C.; Citarella, R. Combined static-cyclic multi-axial crack propagation in cruciform specimens. *Int. J. Fatigue* **2019**, *123*, 296–307. [CrossRef]

10. Bonniot, T.; Doquet, V.; Mai, S.H. Mixed mode II and III fatigue crack growth in a rail steel. *Int. J. Fatigue* **2018**, *115*, 42–52. [CrossRef]

11. Akama, M.; Kiuchi, A. Long co-planar mode III fatigue crack growth under non-proportional mixed mode loading in rail steel. *Proc. Inst. Mech. Eng. Part F J. Rail Rapid Transit* **2012**, *226*, 489–500. [CrossRef]

12. Akama, M.; Kiuchi, A. Fatigue crack growth under non-proportional mixed mode I/III loading in rail and wheel steel. *Tetsu-to-Hagané* **2018**, *104*, 77–86. (In Japanese) [CrossRef]

13. Tada, H.; Paris, P.; Irwin, G. *The Stress Analysis of Cracks Handbook*; Del Research Corporation: Hellertown, PA, USA, 1985.

14. Otsuka, A.; Mori, K.; Miyata, T. The condition of fatigue crack growth in mixed mode loading. *Eng. Fract. Mech.* **1975**, *7*, 429–499. [CrossRef]

15. Pook, L.P. The fatigue crack direction and threshold behavior of mild steel under mixed mode I and III loading. *Int. J. Fatigue* **1985**, *7*, 21–30. [CrossRef]

16. Akama, M.; Kiuchi, A. Fatigue crack propagation tests of rail steel under mixed mode I + III loading. *CAMP-ISIJ* **2010**, *23*, 1169. (In Japanese)

17. Fujii, Y.; Maeda, K.; Otsuka, A. A new test method for mode II fatigue crack growth in hard materials. *J. Soc. Mater. Sci. Jpn.* **2001**, *50*, 1108–1113. (In Japanese) [CrossRef]

18. Wong, S.L.; Bold, P.E.; Brown, M.W.; Allen, R.J. Fatigue crack growth rates under sequential mixed-mode I and II loading cycles. *Fatigue Fract. Eng. Mater. Struct.* **2000**, *23*, 667–674. [CrossRef]

19. Akama, M.; Susuki, I. Fatigue crack growth under mixed mode loading in wheel and rail steel. *Tetsu-to-Hagané* **2007**, *93*, 607–613. (In Japanese) [CrossRef]

20. Lemaitre, J.; Chaboche, J.L. *Mechanics of Solid Materials*; Cambridge University Press: Cambridge, UK, 1990.

21. Dahlin, P.; Olsson, M. The effect of plasticity on incipient mixed-mode fatigue crack growth. *Fatigue Fract. Eng. Mater. Struct.* **2003**, *26*, 577–588. [CrossRef]

22. Schöllmann, M.; Richard, H.A.; Kullmer, G.; Fulland, M. A new criterion for the prediction of crack development in multiaxially loaded structures. *Int. J. Fracture* **2002**, *117*, 129–141. [CrossRef]

23. Hourlier, F.; Pineau, A. Fatigue crack propagation behaviour under complex mode loading. In *Advances in Fracture Research, Proceedings of the Fifth International Conference on Fracture*; Francois, D., Ed.; Pergamon Press: Oxford, UK, 1982; pp. 1833–1840.

24. Murakami, Y.; Takahashi, K.; Kusumoto, R. Threshold and growth mechanism of fatigue cracks under mode II and III loadings. *Fatigue Fract. Eng. Mater. Struct.* **2003**, *26*, 523–531. [CrossRef]

25. Akama, M. Fatigue crack growth under non-proportional mixed mode loading in rail and wheel steel Part 1: Sequential mode I and mode III loading. Accepted for publication in Applied Sciences. *Appl. Sci.* **2019**, *9*, 2006. [CrossRef]

 *applied sciences*

Article

Study of Mixed-Mode Cracking of Dovetail Root of an Aero-Engine Blade Like Structure

Giacomo Canale [1], Moustafa Kinawy [1,*], Angelo Maligno [1], Prabhakar Sathujoda [2] and Roberto Citarella [3]

[1]   Institute of Innovative Sustainable Energy, University of Derby, Kedleston Road, Derby DE1 3HD, UK
[2]   Department of Mechanical and Aerospace Engineering, Bennett University,
      Greater Noida 201310, Uttar Pradesh, India
[3]   Department of Industrial Engineering, University of Salerno, 84084 Salerno, Italy
*   Correspondence: M.Kinawy@derby.ac.uk

Received: 19 July 2019; Accepted: 30 August 2019; Published: 12 September 2019

Abstract: Aerospace structures must be designed in such a way so as to be able to withstand even more flight cycles and/or increased loads. Damage tolerance analysis could be exploited more and more to study, understand, and calculate the residual life of a component when a crack occurs in service. In this paper, the authors are presenting the results of a systematic crack propagation analysis campaign performed on a compressor-blade-like structure. The point of novelty is that different blade design parameters are varied and explored in order to investigate how the crack propagation rate in low cycle fatigue (LCF, at R ratio R = 0) could be reduced. The design parameters/variables studied in this work are: (1) The length of the contact surfaces between the dovetail root and the disc and (2) their inclination angle (denoted as "flank angle" in the aero-engine industry). Effects of the friction coefficient between the disc and the blade root have also been investigated. The LCF crack propagation analyses have been performed by recalculating the stress field as a function of the crack propagation by using the FRacture ANalysis Code (Franc3D®).

Keywords: LCF; crack propagation; blade-disc-Franc3D; mixed-mode cracking

1. Introduction

Fans, compressors, and turbine blades of aero-engines are highly stressed components, especially in the contact area between the blade root and the disc slot [1]. This is due both to the high rotational speed of the shaft and to the aerodynamic load. Consequently, cracks may appear both in low cycle fatigue (LCF, defined in the aero-engine industry as the fatigue caused by the application and release of the main centrifugal and aerodynamic load, with stress ratio R = 0) and high cycle fatigue (HCF, defined in the aero-engine industry as the fatigue induced by the vibration loads). Crack propagation has already been studied in fan, compressor, or turbine blades [1–7]. In recent times, a powerful and simple tool for crack propagation analyses has been released and used in several academic and industrial works: Franc3D [8,9]. It is a state-of-the-art 3D crack propagation tool developed by Cornell University. Franc3D has proven effective in a variety of engineering problems of crack propagation [10–13].

2. Problem Description

The aim of this work is to study the effect on crack propagation of two important design variables of the geometry of the dovetail root in a compressor blade-like structure like the one shown in Figure 1.

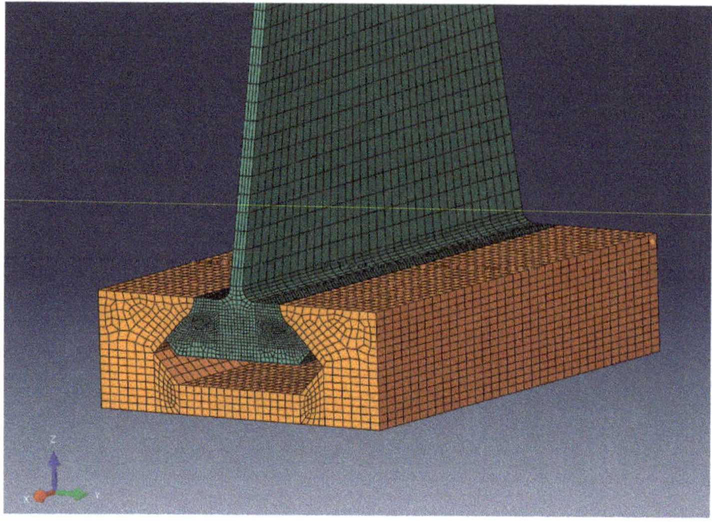

Figure 1. A dovetail joint between the compressor-blade-like structure and the disc representative geometry.

The design variables are shown in Figure 2, on a cross section sketch of the 3D root. They are:

- The length of the contact flank (Figure 2).
- The flank angle ϑ (Figure 2).

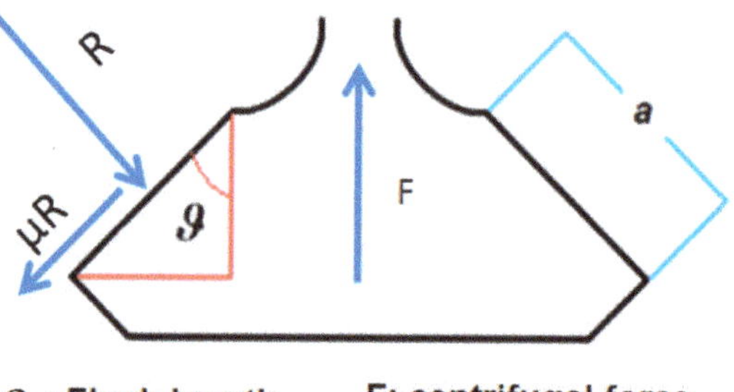

Figure 2. The flank length and flank angle, the design variables in this study.

Three flank design angles were investigated: 30°, 45°, and 60°. For each of these flank angles, two flank lengths were investigated. The short flank was 6.5 mm whilst the long flank was 13 mm. For only one specific case (each analysis takes around 1 week to run on a 12 processors PC), the 45° model with a short flank, the effect of two different friction coefficients (0.1 and 0.7) was studied in order to understand the difference between a perfect frictionless contact and the contact between two worn surfaces.

For the same geometry, two different crack propagation criteria were investigated: A crack driven by an opening mode K_I and a crack driven by mixed-mode represented by K_{eq} [14]. All the crack propagation analyses were performed under LCF, with a ratio of R = 0. It is important to remark that in the aero-engine industry, LCF is intended as the cyclic application of the major load (centrifugal force and aerodynamic pressure in the case of a blade) and its release. The R-ratio is 0. In other words, in LCF, no vibration is taken into account. Linear elastic fracture mechanics can still be applied as the material is far from yielding, with the only exception of a small region close to the crack tip. Given that R = 0, the notation "K" or "ΔK" is therefore used equivalently in this paper.

The initial flaw was assumed as a corner crack. This option was given by Barlow and Chandra in Barlow et al. [1]. The other option to be used, as discussed in [1], would be a semi-elliptic side crack. In reference [1], experimental evidence of these options was also provided.

In this paper, the authors chose the corner crack as the stress field calculated with a finite element model which gave its peak towards the edge of the contact. A crack is more likely to start there.

3. Finite Element (FE) Model of the Blade and Disc Slot

An overview of the blade-like structure used in this assessment is given in Figure 1. In reality, for the purpose of this study, six different Abaqus®FE models were prepared and they are listed in Table 1. All the models have most dimensions in common. Only the angle of the root and the flank length vary from one model to the other. The airfoil thickness (t = 2 mm), root slot width (s = 14 mm), blade axial length (l = 70 mm), and airfoil height (h = 150 mm) are kept constant (Figures 3 and 4). The original angle of twist (undeformed shape) is 30°.

Table 1. Models prepared for the study presented in this paper.

Model Name	Flank Angle (θ)	Flank Length (a) mm
45_short	45°	6
45_long	45°	11
30_short	30°	6
30_long	30°	11
60_short	60°	6
60_long	60°	11

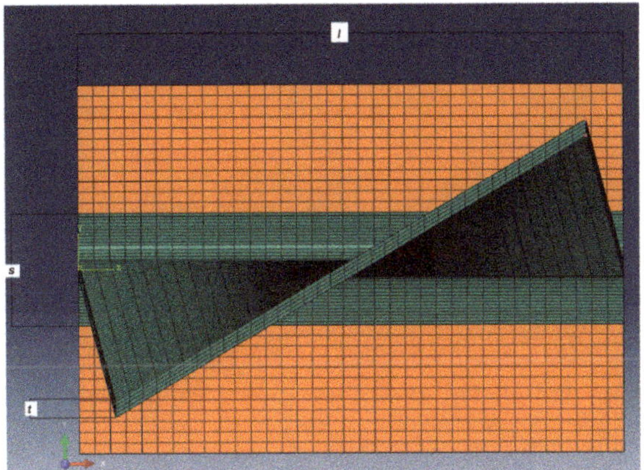

Figure 3. Non-variable dimensions in each of the six FE models.

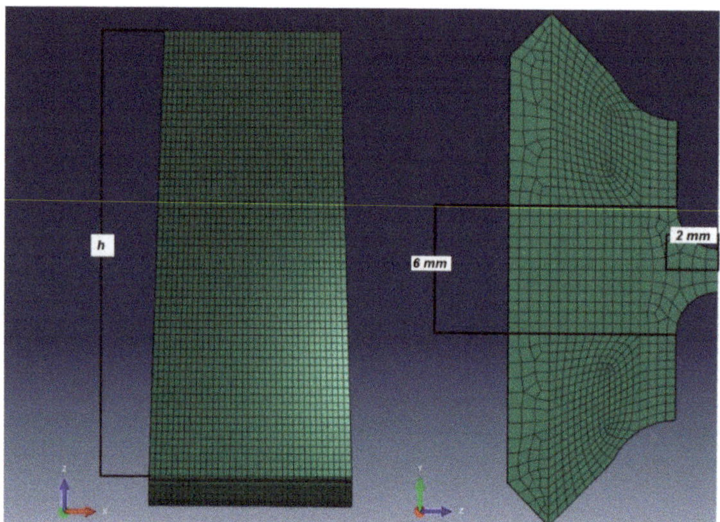

Figure 4. Airfoil length and transition area details.

The blade and disc-like FE models were originally meshed with HEX20 (20 nodes bricks, quadratic elements). Each model was meshed with c.a. 80,000 elements. Static analyses were run before starting the crack propagation study to check the model convergence. It is important to remark that a sub-model of the root was re-meshed by Franc3D itself when the crack was propagating. Franc3D uses TET10 elements (10 nodes tetrahedral elements). An image of the re-meshing is shown in Figure 5.

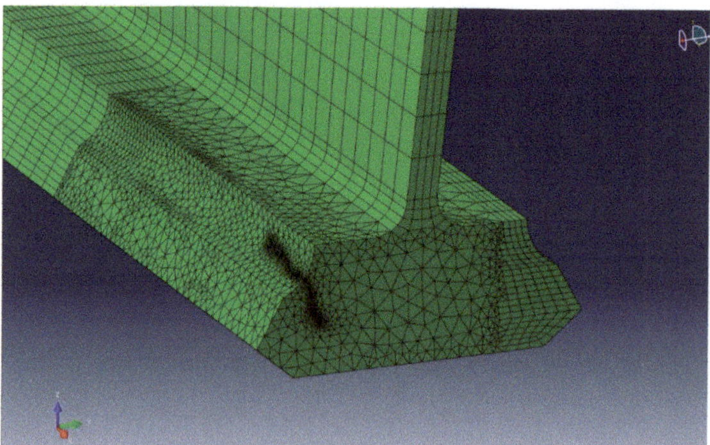

Figure 5. Sub-model containing the crack. The mesh is automatically generated within Franc3D at each iteration.

The blade was made from Ti-6Al-4V. Forging is the manufacturing process used for a small blade like the one studied in reference [15]. For bigger size blades, like compressor rotors of a big turbofan aero-engine, forging can be preceded by a machining operation [16]. Even if the structure is forged, isotropic material properties were assumed for the work proposed in this paper.

3.1. Loads and Boundary Conditions

The base of the block representing the disc is fixed in all the directions as shown in Figure 6. The blade was not restrained in any way and was free to move. Only a surface-to-surface contact was defined along the flanks at the interface of the disc/blade (Figure 7). A "hard" contact between the blade and the block was used with a friction coefficient of 0.7. A similar approach has already been used by Ma et al. [17]. Three different loads are applied to the structure. They are given in Table 2.

Table 2. Loads applied to all the models.

Shaft Rotation Speed (RPM)	Pressure Side (Pressure Value, MPa)	Suction Side (Pressure Value, MPa)
9550	0.18	0.16

The authors are fully aware that real loads may be slightly higher than those really measured in a real blade of a large turbofan. However, this does not affect the generality of the procedure and the results of the design.

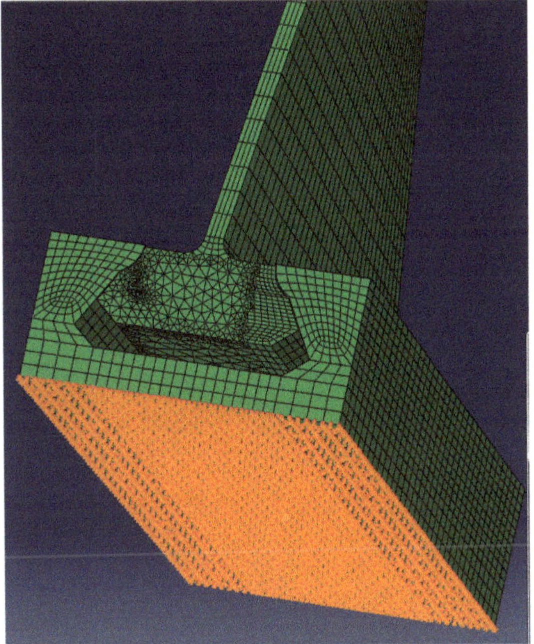

Figure 6. Boundary conditions at the bottom surface of the disc block.

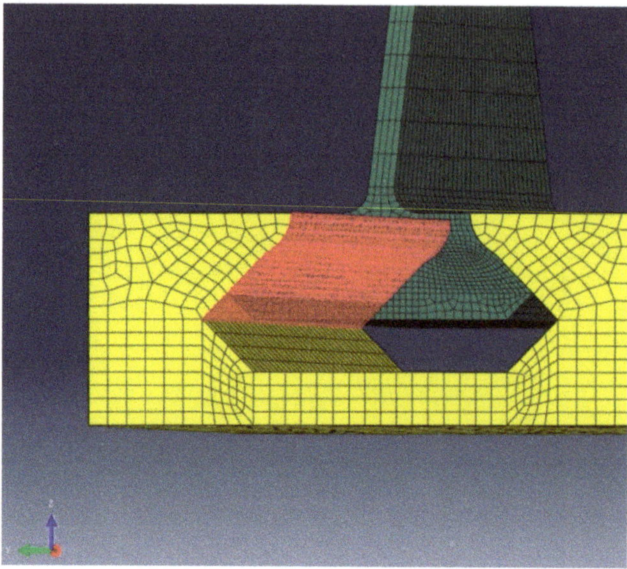

Figure 7. Surface to surface contact definition at the root flanks.

3.2. Material Properties and Crack Propagation Law

The material properties were taken from the work of Al-Emrani et al. [11]. The elastic material properties are given in Table 3. All the analyses were performed assuming a metal temperature = 20 °C. This is quite far from reality, as a gas turbine blade generally works at much hotter conditions. The high temperature will influence material parameters such as Young's modulus and toughness and will also change the friction coefficient at the blade-disc interface. However, for the sake of simplicity, the ambient temperature was assumed as this assumption does not affect the generality of the design procedure.

Table 3. Elastic properties of Ti-6Al-4V used in this paper.

E (MPa)	v	ρ (Mg/mm^3)
115,000	0.33	4.43×10^{-9}

A Paris law was used to calculate the number of LCF cycles. The Paris law reported by Richard and Sander [13] in his book is written in the form:

$$\frac{da}{dN} = C\Delta K^n \tag{1}$$

where:

- a is the crack length
- N is the number of LCF cycles
- C, n are constants
- ΔK is the stress intensity factor range between the maximum and minimum stress field (equals to zero in case of LCF).

The crack propagation data used in the analysis are given in Table 4, and they are all taken from Reference [1], in which K_{IC}, the reported value of fracture toughness, is 1739 MPa·mm$^{0.5}$. For this paper, however, a design value of 869.5 MPa mm$^{0.5}$ was conservatively assumed. In other words, a safety factor of 2 was used for design purposes.

Table 4. Crack propagation (Paris law) for the Ti-6Al-4Va used in this paper (data adapted from [1]).

K_{IC} [MPa mm$^{0.5}$]	da/dN Parameter C	da/dN Parameter n	ΔK_{th} [MPa mm$^{0.5}$]
869.5	1.77×10^{-14}	3.667	121.6

It is quite interesting to compare the relevant crack propagation data from different sources of the open literature. Three examples, for values of R not greater than 0.1, are given in Table 5. In these three examples, the Paris law coefficients have been calculated for a law of the mathematical form as per Equation (1).

Table 5. Crack propagation parameters data for Ti-6Al-4V reported in different literature. Paris Coefficient (C) units *are compatible with (da/dN) in mm/cycle.

Study	K_{IC} [MPa mm$^{0.5}$]	da/dN Parameter C	da/dN Parameter n	ΔK_{th} [MPa mm$^{0.5}$]
Barlow [1]	1739	1.7700×10^{-14}	3.667	121.6
Zhu [18]	1900	0.5085×10^{-14}	3.14	163.8
Ritchie [19]	800	4.8340×10^{-14}	2.5	145.46

The reader can appreciate how different the parameters can be. A set of ad-hoc experiments are therefore needed for the specific engineering application.

In order to predict the number of cycles from the K_{th} to the unstable propagation of the crack (K_{IC}), the Paris law (Equation (1)) has been integrated (clearly, the Paris law is not applicable to the whole range from K_{th} to Kc so the crack length range was such that the near threshold and the near fracture parts of the da/dN-ΔK curve was not involved).

4. Franc3D Simulations Set-Up

Adequate sub-models have been produced from the original Abaqus model. An initial crack, semi-circular, with a radius = 0.5 mm, was inserted c.a. 1.4 mm above the edge of bedding. The initial crack position and orientation (perpendicular to the surface), shown in Figure 8, were kept consistent in all the analysis performed.

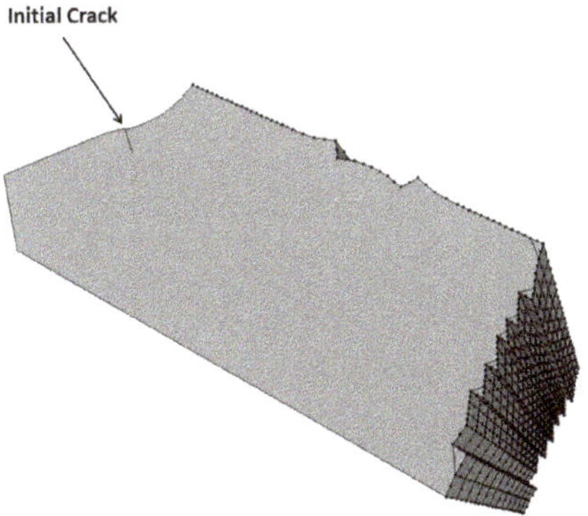

Figure 8. Initial flaw position.

A quasi-static crack propagation analysis was performed. Dynamic effects such as the wave propagation were not taken into account. The "maximum fracture energy" was chosen as an extension criterion [20]. A kink angle perpendicular to the maximum principal stress was chosen. K_{eq} from [14] has been considered equal to K_I. This assumption is valid as long as the values of K_{II} and K_{III} are small compared to the stress intensity in mode I. The validity of this assumption will be discussed in Section 5 when discussing the results. A comparison between K_I and K_{eq} was also performed for the 45° flank angle with short flank length geometry.

Franc3D uses a displacement correlation technique [21] to calculate the stress intensity factors. The displacements on the crack surface (opening, sliding, and tearing) are directly related to the three different stress intensity factors. Stress intensity factors calculated with Franc3D have been extensively validated with analytical solutions in some works available in the open literature [22].

5. Results and Discussion

For each of the analysis performed, the crack evolution was studied. The stress intensity factors K_I, K_{II}, and K_{III}, were plotted as a function of the crack length. These values of the stress intensity factors were plotted along different crack paths. In all the performed analyses, the crack has two directions of propagation, the main leg, along the axial direction of the blade (and of the engine) and the secondary leg, towards the base of the root. An image of general crack propagation is given in Figure 9.

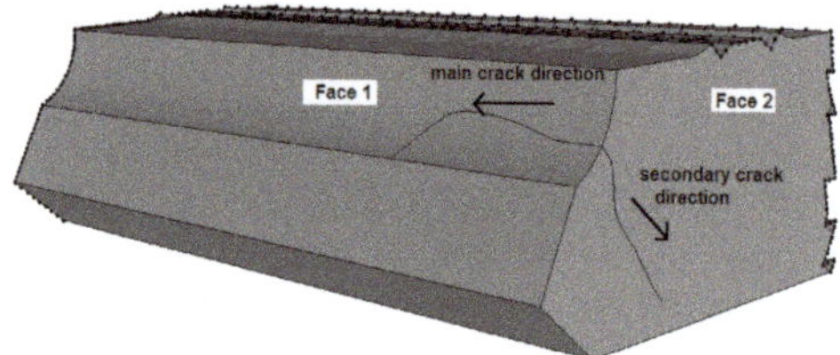

Figure 9. Two leading crack front directions.

All the results presented are given along three specific crack paths named:

- 5% of the crack front (in other words, very close to Face 2 shown in Figure 7).
- 50% of the crack front.
- 95% of the crack front (in other words, the stress intensity is calculated along a crack line very close to Face 1 of Figure 7, along the main crack direction).

A graphic illustration of this crack front is given in Figure 10.

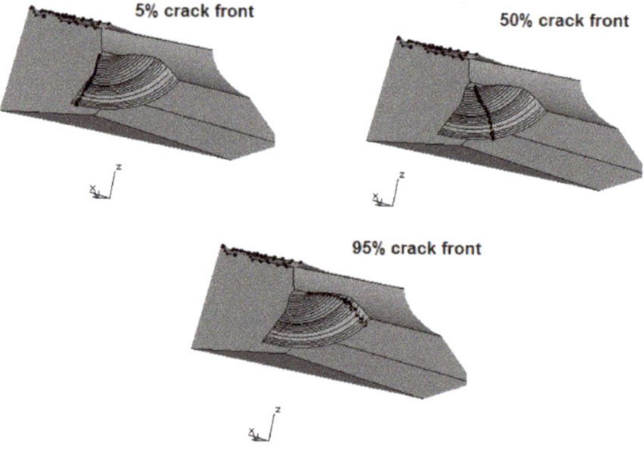

Figure 10. Three crack paths where the stress intensity was computed.

The stress intensity is given in MPa mm$^{0.5}$ whilst the crack length is given in mm.

5.1. The 30° Flank Angle

Two different flank lengths were investigated as design variables when studying the 30° flank angle, a short flank and a long flank, as per the values given in Table 1. The results of K_I, K_{II}, and K_{III} as a function of the crack length are presented for the three different crack fronts. The crack shape of the short flank angle is shown in Figure 11.

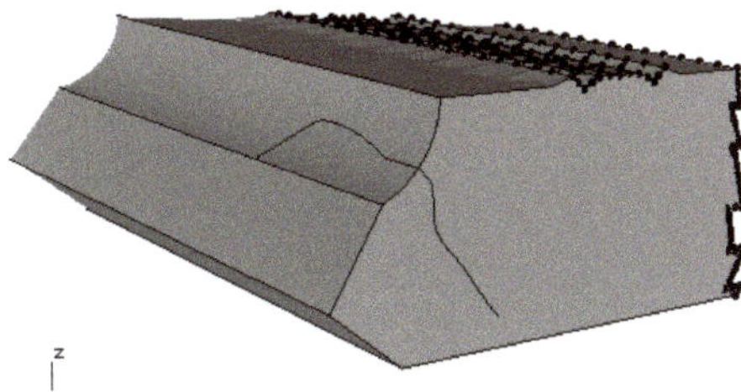

Figure 11. The 30° angle, short flank crack shape.

The stress intensity values along the crack fronts are shown in Figure 12.

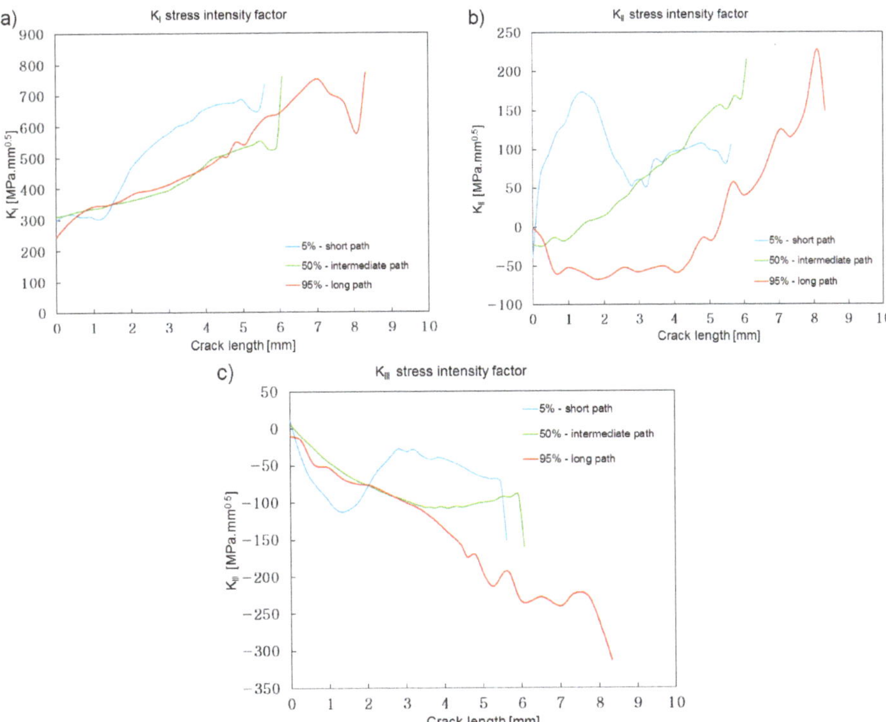

Figure 12. The 30° angle, short flank stress intensity along the three crack paths selected: (**a**) K_I; (**b**) K_{II}; (**c**) K_{III}.

K_I is the biggest contributor to the crack propagation. K_{II}, however, has no negligible values, especially at large crack lengths (c.a. 8 mm crack length). This suggests that the analysed geometry may not be strictly dominated by mode I opening. An analogous consideration can be made for K_{III}, whose value is not negligible compared to K_I, especially when the crack becomes relatively large. This trend is more accentuated when the 95% crack front is considered. The kinks in the plots (Figure 12) towards the end of the simulation (when the crack is large) indicate the crack has reached the edge of bedding, and therefore hitting a numeric singularity. The simulation was therefore stopped. Similar results were obtained when the long flank geometry was analysed. The crack shape obtained with the simulation of the long flank is shown in Figure 13. The stress intensity factors are shown in Figure 14.

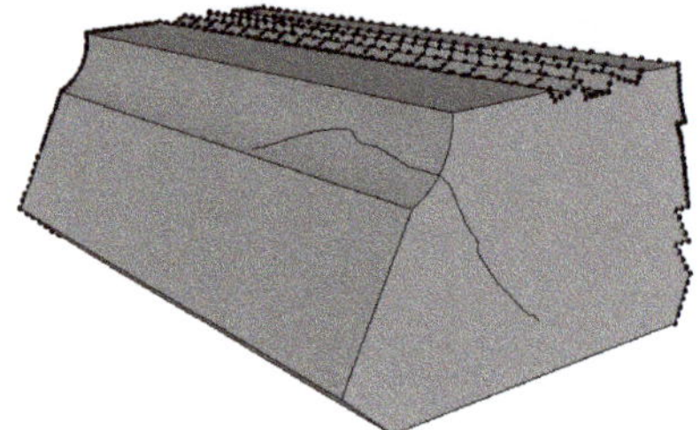

Figure 13. The 30° angle, long flank crack shape.

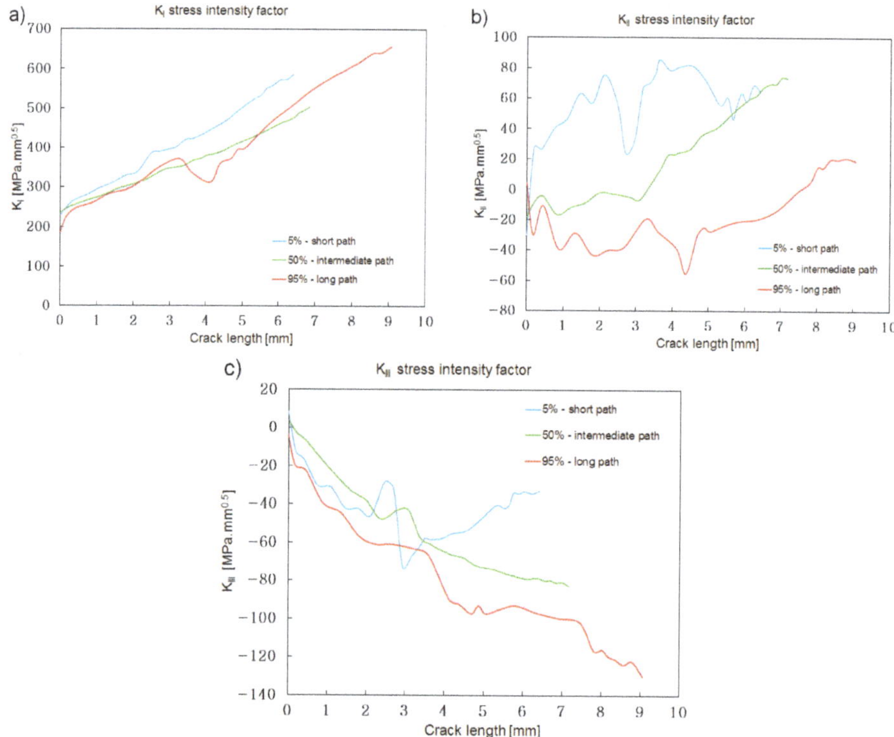

Figure 14. The 30° angle, long flank. Stress intensity along the three crack paths selected: (**a**) K_I; (**b**) K_{II}; (**c**) K_{III}.

In this case, on the other hand, the contribution of K_{II} to the crack propagation is negligible. The contribution of K_{III}, as per the 30° short flange geometry, starts to be significantly important when the crack becomes longer. The crack is therefore not only dominated by opening mode I, but also by a strong shear component, as was suggested by the authors of [11].

5.2. The 60° Flank Angle

Short and long flank lengths have been studied also for the 60° flank angle variant. The final crack shape of the 60° short flank angle is shown in Figure 15.

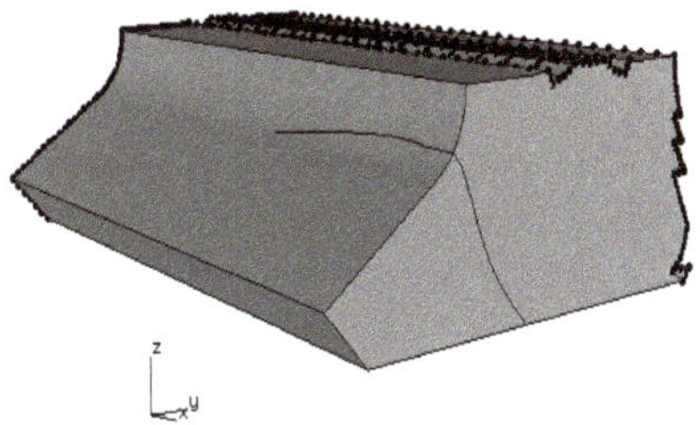

Figure 15. The 60° angle, short flank crack shape.

The stress intensities are given in Figure 16.

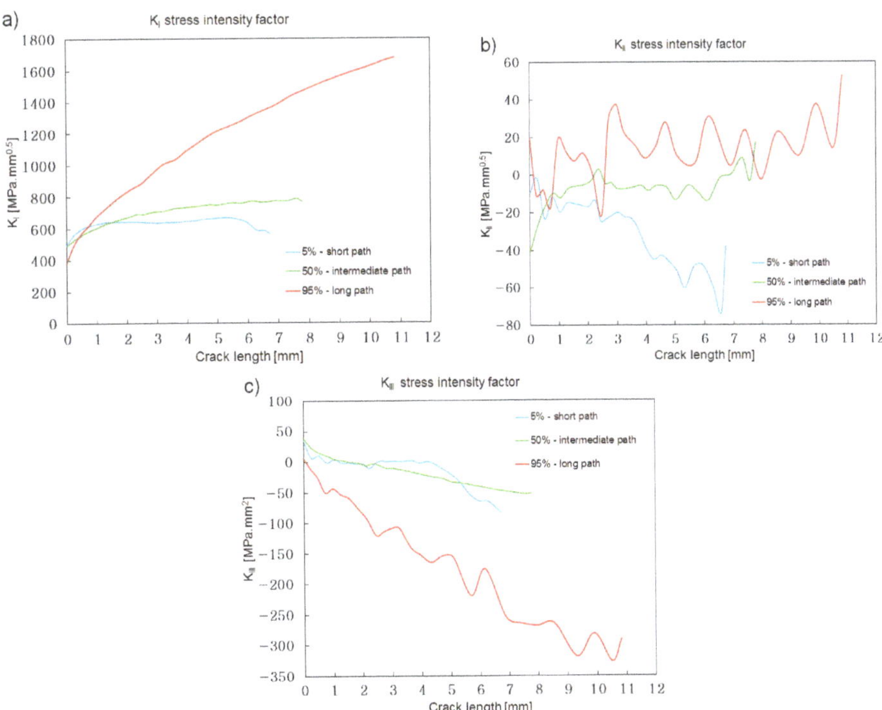

Figure 16. The 60° angle short flank. Stress intensity along the three crack paths selected: (**a**) K_I; (**b**) K_{II}; (**c**) K_{III}.

It can be easily noted that also in this case, the stress intensity drove the crack more rapidly towards an unstable growth located at 95% of the crack front (the portion of the crack closer to the edge of the dovetail contact). The contribution of K_{II} is negligible, whilst for the "long" crack, i.e., when the crack becomes longer, the contribution of K_{III} is quite important.

The crack shape of the long flank geometry option is shown in Figure 17. The stress intensities for the same geometry are given in Figure 18.

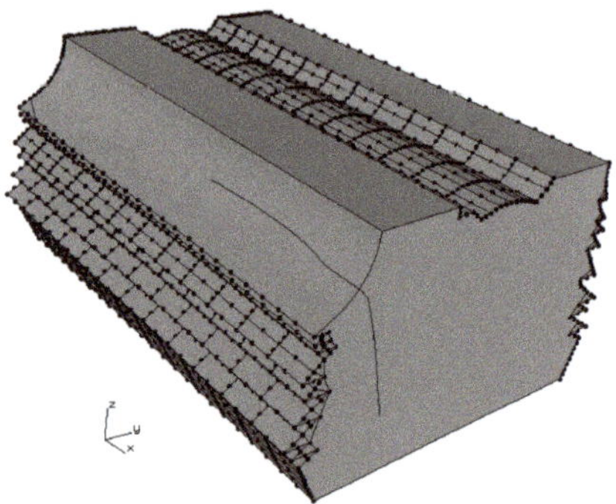

Figure 17. The 60°, long flank crack shape.

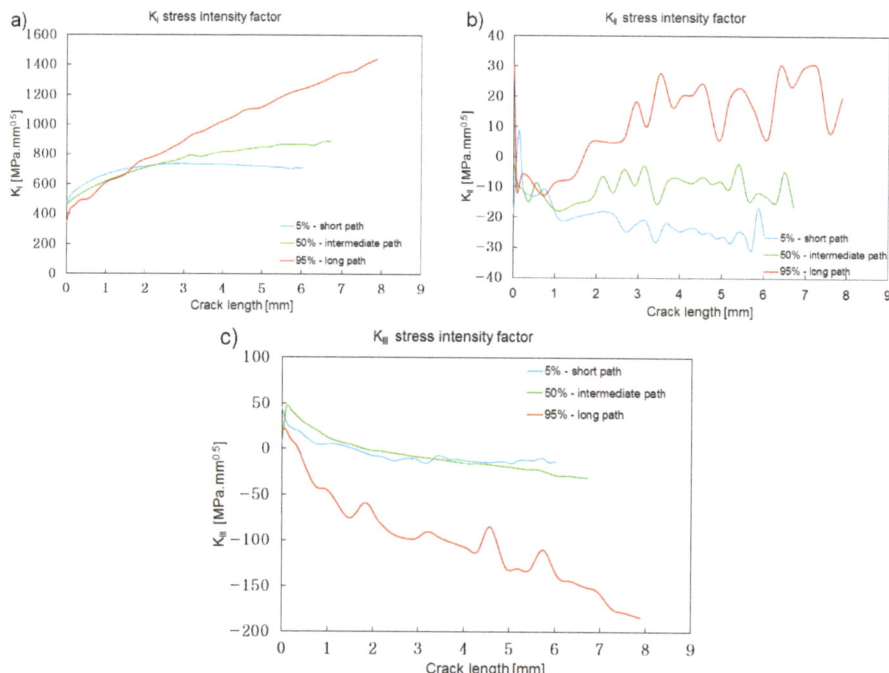

Figure 18. The 60° long flank. Stress intensity along the three crack paths selected: (**a**) K_I; (**b**) K_{II}; (**c**) K_{III}.

5.3. The 45° Flank Angle

The crack shape of the 45° short flank angle is shown in Figure 19.

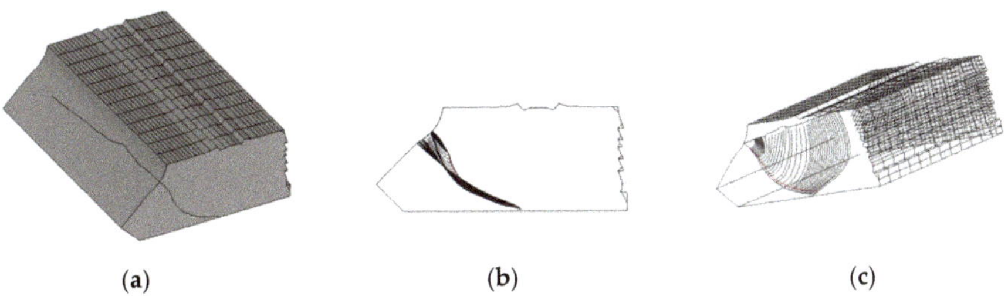

(a)   **(b)**   **(c)**

Figure 19. The 45° angle, short flank crack shape. (**a**) External top view of the crack extension; (**b**) Front view of the crack extension; (**c**) isometric view.

The stress intensities are shown in Figure 20. Also, for this geometry, the critical crack path is along the direction of the engine axis (95% of the crack front). There is a pronounced K_{II} contribution at 5% and 50% of the crack front. Whereas K_{III} has a big influence at the 95% crack front growth.

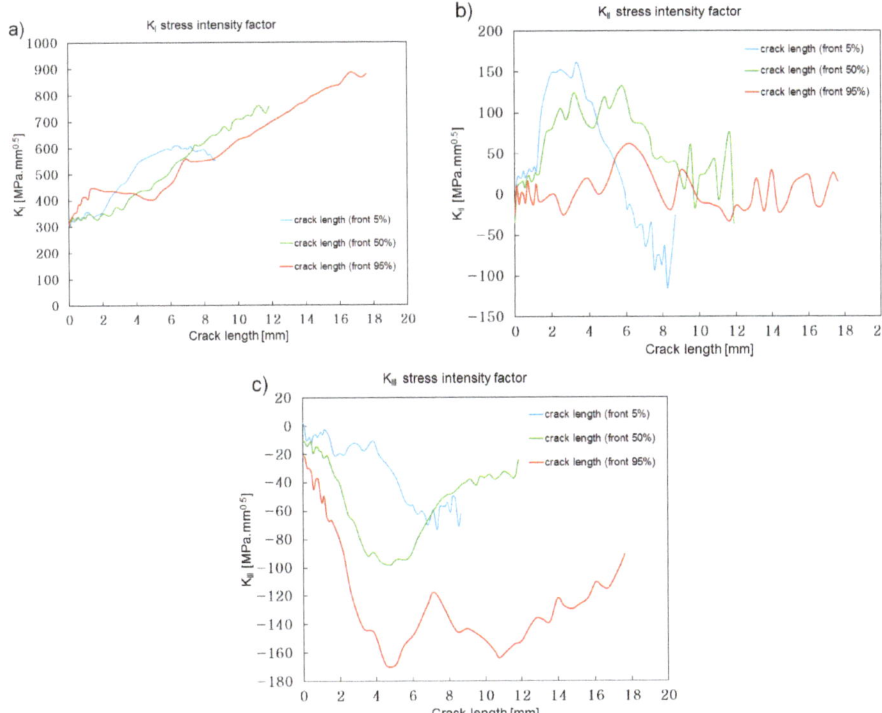

Figure 20. The 45° angle, short flank. Stress intensity along the three crack paths selected: (**a**) K_I; (**b**) K_{II}; (**c**) K_{III}.

Analogously to the other geometries, the long flank of the 45° option was also analysed. The stress intensities are given in Figure 21.

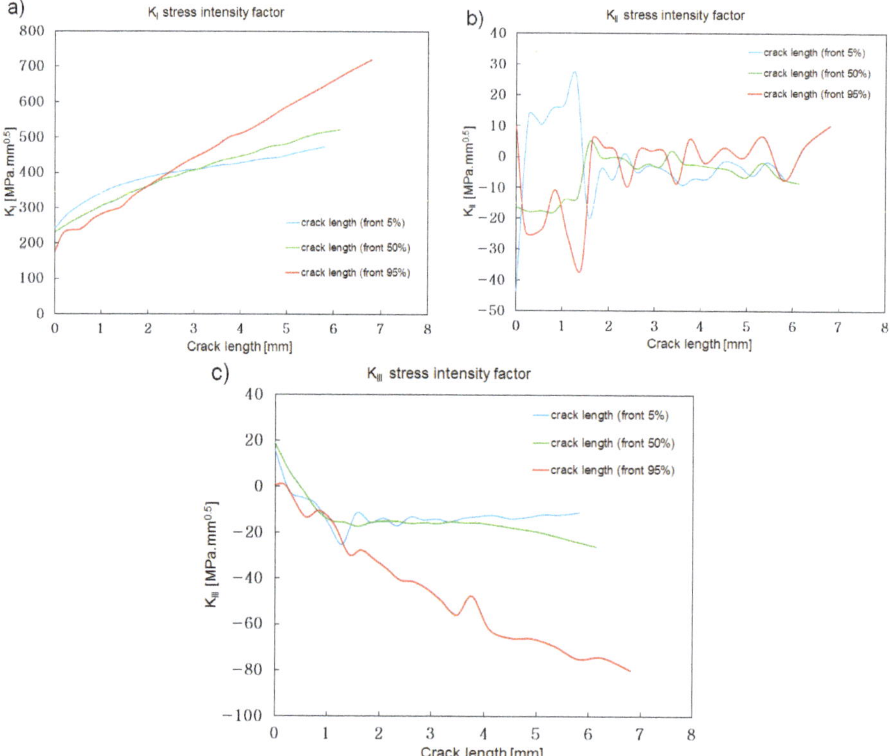

Figure 21. The 45° angle, long flank. Stress intensity along the three crack paths selected: (**a**) K_I; (**b**) K_{II}; (**c**) K_{III}.

It can be noticed that most critical crack path is at 95% of the crack front. After a certain crack length, effects of K_{III} start playing a major role on crack propagation.

5.4. Effects of the Flank Angle: A Comparison

In this section, a comparison of the results of different flank angles is given. The comparison is always performed at the 95% of the crack front. This is to avoid the solution singularity at the free surface of the crack that results from the Franc3D algorithm. The results of the short flank length are shown in Figure 22.

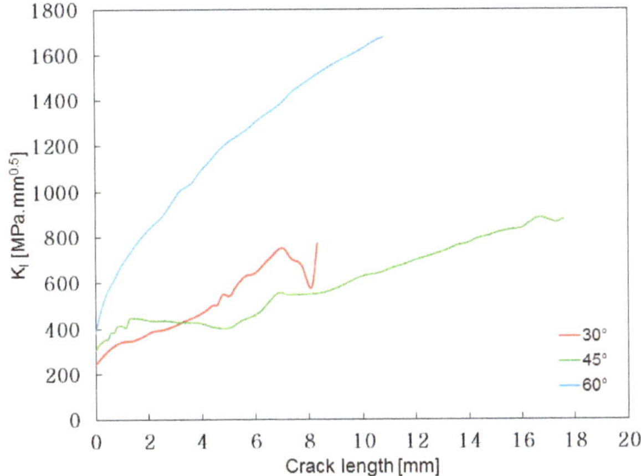

Figure 22. Comparison of the different flank angles at 95% crack front for the short flank design option: KI stress intensity factor.

At small crack sizes, the K_I solution of the 45° angle flank blade had an intermediate trend between the other two angle flank designs. However, the crack changed its growth direction and K_I trend became lower than the other flank angle designs, at a crack size bigger than 3 mm. This is thought to be related to the selected blade geometric parameters. i.e., this behavior could be specific to blade design. This, however, requires more simulations or testing to validate this assumption.

The analogous results for the long flank design options are given in Figure 23. The trend is analogous to the one observed for the short flank option with the only difference 45° angle starts to be beneficial for a longer crack length (c.a. 6 mm).

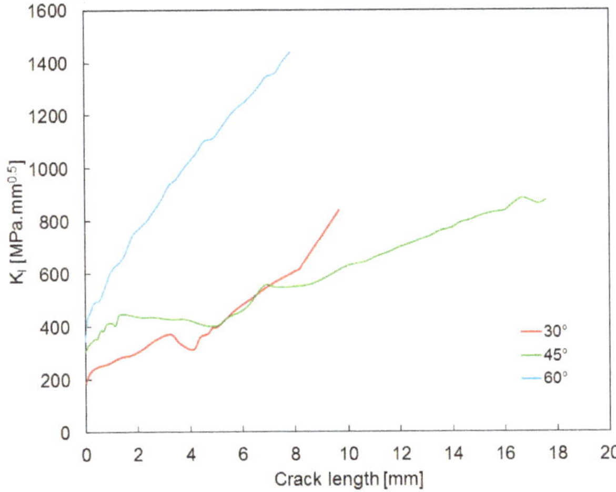

Figure 23. Comparison of the different flank angles at 95% crack front for the long flank design option: K_I stress intensity factor.

The main conclusion of this comparison is that: the 45° angle is beneficial because it helps reduce the stress intensity factor KI when the crack approaches critical values closer to the fracture toughness.

5.5. Effects of the Flank Contact Length: A Comparison

A comparison of all the design angles was made in terms of long against short flank options. It must be remarked that the short flank is preferable in terms of weight penalty. In order to consider the possibility of choosing a longer flank design, there must be a justified technical reasoning in terms of crack initiation (due to the stress distribution) and crack propagation, which is the objective of the present study. A comparison of K_I at 95% crack front for all geometries is given in Figure 24. For the 30° and 60° flank angles, the stress intensity variation from having a shorter flank is very small. For the 45° flank angle, the short flank design option is even beneficial.

This is due to the fact that the contact pressure is not uniformly distributed on the contact surface. A peak of stress in the area of the edge of contact exists in any case, independently of the flank length.

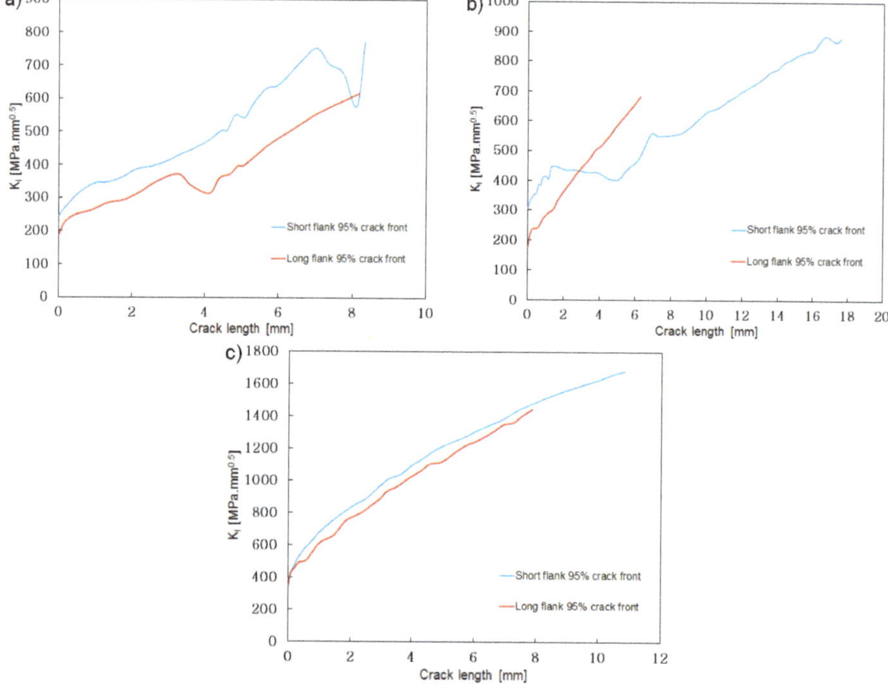

Figure 24. Short vs. long flank for all the flank angle design options: (**a**) 30°; (**b**) 45°; (**c**) 60°.

5.6. Effects of the Friction Coefficient: A Comparison

In order to study the effect of friction, the 45° short flank model was analysed using two different friction coefficients. A friction coefficient $\mu = 0.7$ is useful to simulate a worn flank, whilst a friction coefficient $\mu = 0.1$ is useful to understand the behavior of a perfectly lubricated dovetail contact. At 95% crack front, a back-to-back comparison is shown in Figure 25.

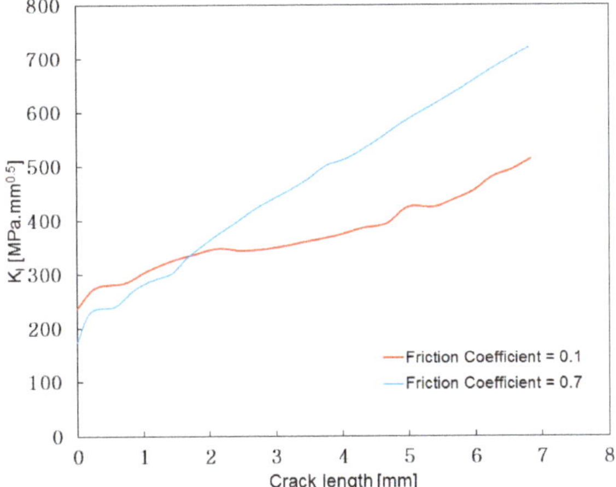

Figure 25. Crack propagation behavior with low and high metal-to-metal friction coefficient. The geometry is a 45° short flank, 95% crack front.

As expected, a lower friction coefficient implies the blade can slide further up the blade flank and hence a reduction of the stress fields and the crack propagation is slower, especially when the crack is growing longer than >2 mm.

An overview of the crack shape difference between the low and high friction is shown in Figure 26.

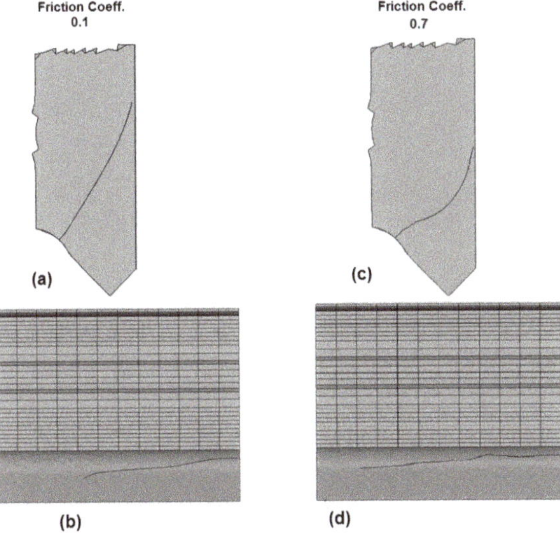

Figure 26. Crack shape and direction. A direct comparison between low and high friction coefficients. The geometry is a 45° short flank. (**a**) Friction coefficient = 0.1 front view; (**b**) Friction coefficient = 0.1 top view; (**c**) Friction coefficient = 0.7 front view (**d**) Friction coefficient = 0.1 top view.

5.7. Effects of K_{eq}

A crack subjected to three-dimensional mixed-mode loading is capable of growth if $\Delta K_{eq} > \Delta K_{th}$. In other words, the crack propagates if a stress intensity factor written as a function of K_I, K_{II}, and K_{III} (and not K_I alone) exceeds the propagation threshold. The Franc3D analysis was repeated by using K_{eq} instead of K_I for the 45° short flank geometry. This numerical simulation is relevant as it was observed that K_{II} and even most importantly K_{III} may play a relevant role in such complex cracking scenarios.

The value of K_{eq} used was taken from reference [14] and it is given in Equation (2).

$$\Delta K_{eq} = \frac{\Delta K_I}{2} + \frac{\sqrt{\Delta K_I^2 + 5.336\,\Delta K_{II}^2 + 4\,\Delta K_{III}^2}}{2} \tag{2}$$

This 3D mixed-mode criterion was compared with other criteria available in the open literature (like Pook and Schoellmann) [23] and very small difference was found in terms of crack propagation (direction, kink, twist angle, and so on).

A comparison between a propagation driven by K_I alone and a propagation driven by K_{eq} is shown in Figure 27. The geometry analysed is the 45° short flank.

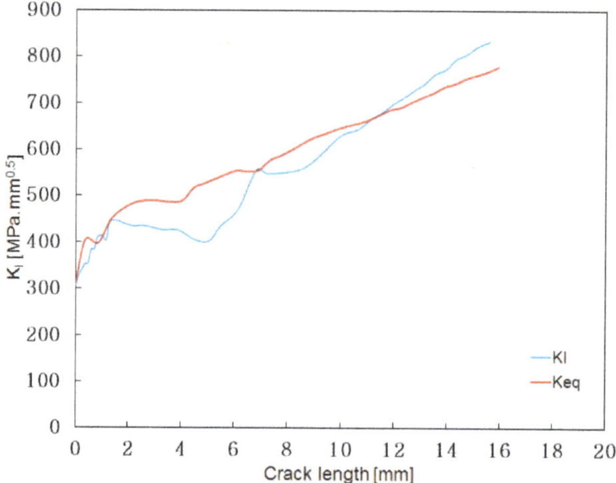

Figure 27. Comparison between a K_I and a K_{eq} driven crack for a 45° short flange geometry, 95% crack front.

It could be clearly observed that the K_{eq} is higher than considering only the K_{IC} failure criterion when the crack is relatively short. After a certain length, however, the crack behaviour remains almost unaffected until 12 mm crack length. Compared to that size, the K_I will have a higher trend than K_{eq}. The different crack shapes are shown in Figure 28. It can be seen that the long path of the crack will deviate in such a way as to be able to follow the edge of the contact.

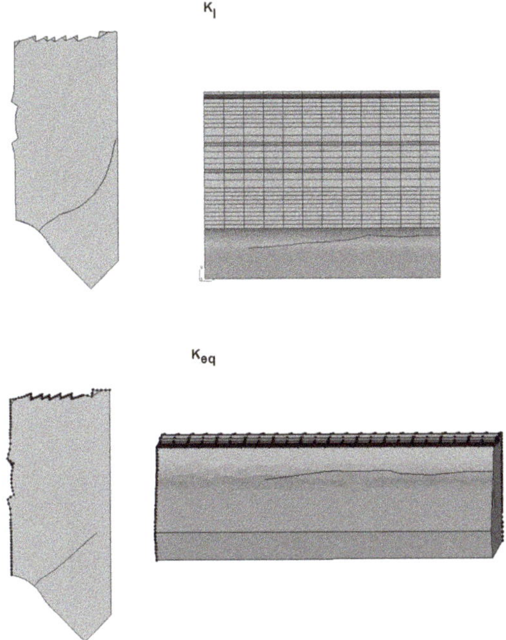

Figure 28. Crack shape and direction, a direct comparison between K$_I$ and K$_{eq}$. The geometry is a 45°
short flank.

5.8. Untwist and Tip Clearances

Understanding the crack progression on a blade root is not only important from the structural
integrity point of view, but also in terms of displacements. The crack progression, in fact, may affect the
displacements of the tip of the blade, both at the leading and trailing edge. Consequently, the untwist
may be affected and even the entire performance of the subsystem it works within or its stability
(surge may be caused by an excess of tip clearance). With the term "untwist" the reduction in the
angle of torsion induced by the centrifugal force of a rotating component in a turbomachinery is meant.
The initial angle of twist φ (undeformed geometry) is 30° for all the geometries analysed (Figure 29).
Such an angle tends to become smaller when the centrifugal load is applied. The variation from this
original angle is the "untwist".

It is also useful to monitor the vertical displacements of point A (the leading edge tip) and point B
(the trailing edge tip). Such a parameter is of particular interest when assessing tip clearances in a
sub-system, such as axial compressors or turbines.

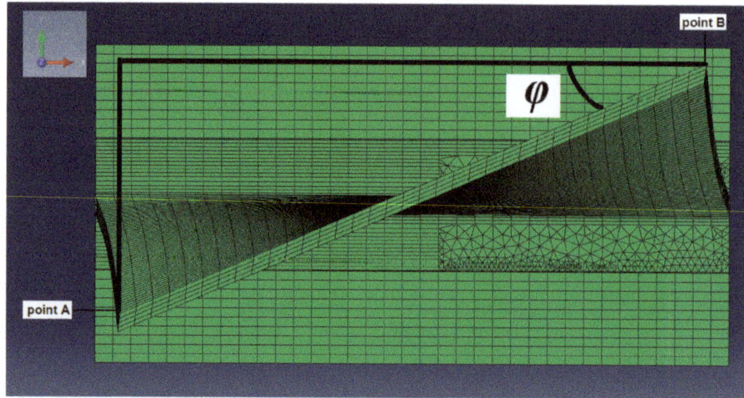

Figure 29. Definition of angle of twist. Its reduction when a centrifugal load is applied is called "untwist".

The results of such a deformation as a function of the crack progression are given in Tables 6 and 7 for the short flanks geometries with 30° and 60° flank angles.

Table 6. Deformation results for the 30° short flank geometry.

Crack Length [mm]	Untwist [Degrees]	LE Vertical Deflection [mm]	TE Vertical Deflection [mm]
0	−7.2218	0.8828	0.7257
1.3738	−7.2202	0.8826	0.726
2.0152	−7.2157	0.8821	0.7268
3.3353	−7.2139	0.8819	0.7276
6.5189	−7.2054	0.8806	0.7297

Table 7. Deformation results for the 60° short flank geometry.

Crack Length [mm]	Untwist [Degrees]	LE Vertical Deflection [mm]	TE Vertical Deflection [mm]
0	−7.1913	0.9823	0.8126
1.262	−7.1874	0.9818	0.8137
2.722	−7.1787	0.9807	0.8163
4.343	−7.1639	0.9788	0.8207
6.886	−7.132	0.9752	0.8299
9.894	−7.0661	0.969	0.8468
10.816	−7.0595	0.9671	0.8524

It can be noted that the deformation was not heavily affected. This happened because the blade root was heavily constrained within the disc slot as shown in Figures 6 and 7. As the crack progressed, a different distribution of stress was observed. The area of contact immediately close to the crack front tends to experience a stress relief whilst the undamaged area of the flank tends to experience a stress increase as shown in Figure 30.

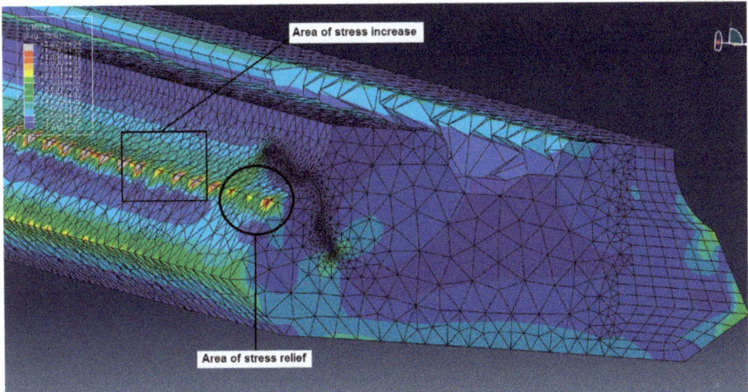

Figure 30. Contact flank stress re-distribution as a function of the crack progression.

It is important to remark that because of the above-mentioned re-distribution, cracks other than the one analysed may be initiated and propagate. This phenomenon has not been taken into account in this work.

5.9. Damage Tolerant Cycles: An Example

A design value of 800 MPa mm$^{0.5}$ was taken as the critical fracture toughness. Equation (1) was used to calculate the damage tolerant life of each design solution. Results are given in Table 8.

Table 8. Damage tolerant life of the analysed geometries.

Geometry	Flank Length	Friction Coefficient	Crack Length at 810 MPa mm$^{0.5}$	Number of Cycles at 810 MPa mm$^{0.5}$
45°	Short	0.7	14.73	17,842
45°	Long	0.7	7.80	17,644
45°	Short	0.1	11.16	17,805
60°	Short	0.7	1.845	16,528
60°	Long	0.7	2.38	16,836
30°	Short	0.7	8.61	17,681
30°	Long	0.7	9.98	17,734

The difference between the residual life of the different geometries is not a significant number. It can however be noticed that the 45° flank with short flank offers the best damage tolerance option. This is also true in terms of crack length reached at the design fracture toughness. The "45° degrees short" can tolerate longer cracks without catastrophic failure.

6. Conclusions

Several blade dovetail root geometries with different flank angles and lengths were analysed with the same structural loads under the same initial flaw, by using Franc3D$^©$ and Abaqus. Effects of friction coefficient and the utilisation of an equivalent K_{eq} calculation method were also studied. The following was concluded:

- The 45° flank angle offers a lower crack propagation rate compared to the 30° and 60° geometries. In other words, the damage tolerant life offered by the 45° solution is better than the other two flank angle options.
- For a given crack length, the stress intensity K_I given by the short and long flank (for given flank angle) is comparable. The short flank has, in some cases, slightly higher values of stress intensity

factors, but a design choice of longer flank will have the disadvantage of increasing the weight of the structure only to obtain a negligible advantage in terms of crack propagation. This advantage would not exist in the 45° design option.

- The untwist of the leading and trailing edge tip displacements are virtually unaffected by the crack propagation. The stress re-distribution along the flank length, however, may induce the nucleation and propagation of different (and numerous) crack fronts. This case has not been considered in this work.

- The reaction, R, and hence the friction force on the flank surface, μR, increase inversely with ϑ, i.e., $R \propto 1/(\cos \vartheta)$. Once the blade is loaded under centrifugal force, the friction at the flank surface will restrain the blade from sliding up against the disc flank and hence more strain energy would be released at the crack front which will extend the crack further towards the blade root. In other terms, the higher the friction, the faster the crack would propagate.

- A low friction coefficient (given for example by a lubricated contact) is beneficial in slowing down the crack propagation, even if its influence is not as massive as expected.

- Setting the failure criterion in Franc3D to K_I will have no significant difference from using K_{eq} to calculate the K at crack sizes below 12 mm as the contribution of other modes in the K_{eq} equation is not significant. However, more simulations would be required to confirm this trend for higher crack sizes and blade geometries.

- No influence of vibration was studied. This will be part of future investigations.

Author Contributions: Formal analysis, G.C. and M.K.; Visualization, A.M. and P.S.; Writing—original draft, G.C. and M.K.; Writing—review & editing, A.M., P.S. and R.C.

Funding: This research received no external funding.

Acknowledgments: The authors would like to thank John Schofield, John Bolger, Richard Coghlan (Rolls-Royce plc, Derby, UK), and Felice Rubino for their ongoing support. Without their contribution this work would not have been possible.

Conflicts of Interest: The authors declare no conflict of interest exists as a result of publishing this work.

References

1. Barlow, K.W.; Chandra, R. Fatigue crack propagation in an aircraft engine fan blade attachment. *Int. J. Fatigue* **2005**, *27*, 1661–1668. [CrossRef]
2. Poursaeidi, E.; Bakthtini, H. Fatigue crack simulation in a first stage of compressor blade. *Eng. Fail. Anal.* **2014**, *45*, 314–325. [CrossRef]
3. Poursaeidi, E.; Salavatien, M. Fatigue crack growth simulation in a generator fan blade. *Eng. Fail. Anal.* **2009**, *16*, 888–898. [CrossRef]
4. Giannella, V.; Vivo, E.; Mazzeo, M.; Citarella, R. FEM-DBEM approach to simulate crack propagation in a turbine vane segment undergoing a fatigue load spectrum. *Procedia Struct. Integr.* **2018**, *12*, 479–491. [CrossRef]
5. Giannella, V.; Perrella, M.; Shlyannikov, V.N. Fatigue crack growth in a compressor stage of a turbofan engine by FEM-DBEM approach. *Procedia Struct. Integr.* **2018**, *12*, 404–415. [CrossRef]
6. Citarella, R.; Giannella, V.; Vivo, E.; Mazzeo, M. FEM-DBEM approach for crack propagation in a low pressure aeroengine turbine vane segment. *Theor. Appl. Fract. Mech.* **2016**, *86*, 143–152. [CrossRef]
7. Giannella, V.; Dhondt, G.; Kontermann, C.; Citarella, R. Combined static-cyclic multi-axial crack propagation in cruciform specimens. *Int. J. Fatigue* **2019**, *123*, 296–307. [CrossRef]
8. Cornell Fracture Group, Cornell University. Available online: www.cfg.cornell.edu (accessed on 11 November 2018).
9. Cornell Fracture Group. *FRANC3D Menu & Dialog Reference*; Cornell University: Ithaca, NY, USA, 2002.
10. Seifi, R.; Omidvar, N. Fatigue crack growth under mixed mode I+III loading. *Mar. Struct.* **2013**, *34*, 1–15. [CrossRef]

11. Aygül, M.; Al-Emrani, M.; Barsoum, Z.; Leander, J. An investigation of distortion-induced fatigue cracking under variable amplitude loading using 3D crack propagation analysis. *Eng. Fail. Anal.* **2014**, *45*, 151–163. [CrossRef]

12. Pettit, R.; Annigeri, B.; Owen, W.; Wawrzynek, P. Next generation 3D Mixed Mode fracture propagation theory including HCF-LCF interaction. *Eng. Fract. Mech.* **2013**, *102*, 1–14. [CrossRef]

13. Giannella, V.; Citarella, R.; Perrella, M.; Shlyannikov, V. Surface crack modelling in an engine compressor disc. *Theor. Appl. Fract. Mech.* **2019**, *103*, 102279. [CrossRef]

14. Richard, H.A.; Sander, M. *Fatigue Crack Growth Detect-Assess-Avoid*; Springer: Berlin/Heidelberg, Germany, 2016.

15. Saigal, A.; Zhan, K.; Chan, T.S. Forging of compressor blades: Temperature and ram velocity effects. *J. Eng. Gas Turbines Power* **1995**, *117*, 528–533. [CrossRef]

16. Meetham, G.W. *The Development of Gas Turbine Materials*; Applied Science Publisher; Springer: Berlin/Heidelberg, Germany, 1981.

17. Ma, H.; Wang, D.; Tai, X.; Wen, B. Vibration response analysis of blade-disc dovetail structure under blade tip rubbing condition. *J. Vib. Control* **2015**, *23*, 252–271. [CrossRef]

18. Zhu, Y.; Xiong, J.; Lv, Z.; Zhao, Y. Testing and Evaluation for fatigue crack propagation of Ti-6Al-4Va/ELI and 7050-T7452 alloys at high temperatures. *Chin. J. Aeronaut.* **2018**, *31*, 1388–1398. [CrossRef]

19. Ritchie, R.O.; Davidson, D.L.; Boyce, B.L.; Campbell, J.P.; Roder, O. High Cycle Fatigue of Ti-6Al-4Va. *Fatigue Fract. Eng. Mater. Struct.* **1999**, *22*, 621–631. [CrossRef]

20. Carter, B.; Wawrzynek, P. *Franc3D Reference Manual*; Version 7; Fracture Analysis Consultants, Inc.: New York, NY, USA, 2016.

21. Ashoaibi, A.M.; Ariffin, A.K. Evaluation of Stress Intensity Factor Using Displacement Correlation Techniques. *J. Kejuruteraan* **2008**, *20*, 75. [CrossRef]

22. Dias, D.E. SIF Calculations for Multistable Damage Using Franc3d. Master's Thesis, University of Bandung, Bandung, Indonesia, 2008.

23. Richard, H.A.; Eberlein, A. Material characteristics at 3D mixed mode loading. In Proceedings of the 21st European Conference on Fracture, ECF21, Catania, Italy, 20–24 June 2016.

MDPI
St. Alban-Anlage 66
4052 Basel
Switzerland
Tel. +41 61 683 77 34
Fax +41 61 302 89 18
www.mdpi.com

Applied Sciences Editorial Office
E-mail: applsci@mdpi.com
www.mdpi.com/journal/applsci